Advances in Information Security

Volume 55

Series Editor

Sushil Jajodia, Center for Secure Information Systems, George Mason University, Fairfax, VA, 22030-4444, USA

For further volumes:
http://www.springer.com/series/5576

Robinson E. Pino
Editor

Network Science and Cybersecurity

 Springer

Editor
Robinson E. Pino
ICF International
Fairfax, VA
USA

ISSN 1568-2633
ISBN 978-1-4899-9065-5   ISBN 978-1-4614-7597-2   (eBook)
DOI 10.1007/978-1-4614-7597-2
Springer New York Heidelberg Dordrecht London

Printed on acid-free paper

Springer is part of Springer Science+Business Media (www.springer.com)

Preface

In terms of network science and cybersecurity, the challenge is the ability to perceive, discover, and prevent malicious actions or events within the network. It is clear that the amount of information traffic data types has been continuously growing which makes the job of a security analyst increasingly difficult and complex. Therefore, it is the goal of this book to offer basic research solutions into promising emerging technologies that will offer and enable enhanced cognitive and high performance capabilities to the human analyst charged with monitoring and securing the network. The work contained herein describes the following research ideas.

Towards Fundamental Science of Cyber Security provides a framework describing commonly used terms like "Science of Cyber" or "Cyber Science" which have been appearing in the literature with growing frequency, and influential organizations initiated research initiatives toward developing such a science even though it is not clearly defined. The chapter offers a simple formalism of the key objects within cyber science and systematically derives a classification of primary problem classes within the science.

Bridging the Semantic Gap—Human Factors in Anomaly-Based Intrusion Detection Systems examines the "semantic gap" with reference to several common building blocks for anomaly-based intrusion detection systems. Also, the chapter describes tree-based structures for rule construction similar to those of modern results in ensemble learning, and suggests how such constructions could be used to generate anomaly-based intrusion detection systems that retain acceptable performance while producing output that is more actionable for human analysts.

Recognizing Unexplained Behavior in Network Traffic presents a framework for evaluating the probability that a sequence of events is not explained by a given a set of models. The authors leverage important properties of this framework to estimate such probabilities efficiently, and design fast algorithms for identifying sequences of events that are unexplained with a probability above a given threshold.

Applying Cognitive Memory to CyberSecurity describes a physical implementation in hardware of neural network algorithms for near- or real-time data mining, sorting, clustering, and segmenting of data to detect and predict criminal behavior using Cognimem's CM1 K cognitive memory as a practical and commercially available example. The authors describe how a vector of various attributes can be constructed, compared, and flagged within predefined limits.

Understanding Cyber Warfare discusses the nature of risks and vulnerabilities and mitigating approaches associated with the digital revolution and the emergence of the World Wide Web. The discussion geared mainly to articulating suggestions for further research rather than detailing a particular method.

Design of Neuromorphic Architectures with Memristors presents the design criteria and challenges to realize Neuromorphic computing architectures using emerging memristor technology. In particular, the authors describe memristor models, synapse circuits, fundamental processing units (neural logic blocks), and hybrid CMOS/memristor neural network (CMHNN) topologies using supervised learning with various benchmarks.

Nanoelectronics and Hardware Security focuses on the utilization of nano-electronic hardware for improved hardware security in emerging nanoelectronic and hybrid CMOS-nanoelectronic processors. Specifically, features such as variability and low power dissipation can be harnessed for side-channel attack mitigation, improved encryption/decryption, and anti-tamper design. Furthermore, the novel behavior of nanoelectronic devices can be harnessed for novel computer architectures that are naturally immune to many conventional cyber attacks. For example, chaos computing utilizes chaotic oscillators in the hardware implementation of a computing system such that operations are inherently chaotic and thus difficult to decipher.

User Classification and Authentication for Mobile Device Based on Gesture Recognition describes a novel user classification and authentication scheme for mobile devices based on continuous gesture recognition. The user's input patterns are collected by the integrated sensors on an Android smartphone. A learning algorithm is developed to uniquely recognize a user during their normal interaction with the device while accommodating hardware and biometric features that are constantly changing. Experimental results demonstrate a great possibility for the gesture-based security scheme to reach sufficient detection accuracy with an undetectable impact on user experience.

Hardware-Based Computational Intelligence for Size, Weight, and Power Constrained Environments examines the pressures pushing the development of unconventional computing designs for size, weight, and power constrained environments and briefly reviews some of the trends that are influencing the development of solid-state neuromorphic systems. The authors also provide high level examples of selected approaches to hardware design and fabrication.

Machine Learning Applied to Cyber Operations investigates machine learning techniques that are currently being researched and are under investigation within the Air Force Research Laboratory. The purpose of the chapter is primarily to educate the reader on some machine learning methods that may prove helpful in cyber operations.

Detecting Kernel Control-flow Modifying Rootkits proposes a Virtual Machine Monitor (VMM)-based framework to detect control-flow modifying kernel rootkits in a guest Virtual Machine (VM) by checking the number of certain hardware events that occur during the execution of a system call. Our technique leverages the Hardware Performance Counters (HPCs) to securely and efficiently count the monitored hardware events. By using HPCs, the checking cost is significantly reduced and the temper-resistance is enhanced.

Formation of Artificial and Natural Intelligence in Big Data Environment discusses Holographic Universe representation of the physical world and its possible corroboration. The author presents a model that captures the cardinal operational feature of employing unconsciousness for Big Data and suggests that models of the brain without certain emergent unconsciousness are inadequate for handling the Big Data situation. The suggested "Big Data" computational model utilizes all the available information in a shrewd manner by manipulating explicitly a small portion of data on top of an implicit context of all other data.

Alert Data Aggregation and Transmission Prioritization over Mobile Networks presents a novel real-time alert aggregation technique and a corresponding dynamic probabilistic model for mobile networks. This model-driven technique collaboratively aggregates alerts in real-time, based on alert correlations, bandwidth allocation, and an optional feedback mechanism. The idea behind the technique is to adaptively manage alert aggregation and transmission for a given bandwidth allocation. This adaptive management allows the prioritization and transmission of aggregated alerts in accordance with their importance.

Semantic Features from Web-traffic Streams describes a method to convert web-traffic textual streams into a set of documents in a corpus to allow use of established linguistic tools for the study of semantics, topic evolution, and token-combination signatures. A novel web-document corpus is also described which represents semantic features from each batch for subsequent analysis. This representation thus allows association of the request string tokens with the resulting content, for consumption by document classification and comparison algorithms.

Concurrent Learning Algorithm and the Importance Map presents machine learning and visualization algorithms developed by the U.S. National Security Agency's Center for Exceptional Computing. The chapter focuses on a cognitive approach and introduces the algorithms developed to make the approach more attractive. The Concurrent Learning Algorithm (CLA) is a biologically inspired algorithm, and requires a brief introduction to neuroscience. Finally, the Importance Map (IMAP) algorithm will be introduced and examples given to clearly illustrate its benefits.

Hardware Accelerated Mining of Domain Knowledge introduces cognitive domain ontologies (CDOs) and examines how they can be transformed into constraint networks for processing on high-performance computer platforms. The constraint networks were solved using a parallelized generate and test exhaustive depth first search algorithm. Two compute platforms for acceleration are examined: Intel Xeon multicore processors, and NVIDIA graphics processors (GPGPUs). The scaling of the algorithm on a high-performance GPGPU cluster achieved estimated speed-ups of over 1,000 times.

Memristors and the Future of Cyber Security Hardware covers three approaches to emulate a memristor-based computer using artificial neural networks and describes how a memristor computer could be used to solve Cyber security problems. The memristor emulation neural network approach was divided into three basic deployment methods: (1) deployment of neural networks on the traditional Von Neumann CPU architecture, (2) software-based algorithms deployed on the Von Neumann architecture utilizing a Graphics Processing Units (GPUs), and (3) a hardware architecture deployed onto a field-programmable gate array.

This book is suitable for engineers, technicians, and researchers in the fields of cyber research, information security and systems engineering, etc. It can also be used as a textbook for senior undergraduate and graduate students. Postgraduate students will also find this a useful sourcebook since it shows the direction of current research. We have been fortunate in attracting outstanding class researchers as contributors and wish to offer our thanks for their support in this project.

Dr. Robinson E. Pino works with ICF International and has expertise within technology development, program management, government, industry, and academia. He advances state-of-the-art cybersecurity solutions by applying autonomous concepts from computational intelligence and neuromorphic computing. Previously, Dr. Pino was a senior electronics engineer at the U.S. Air Force Research Laboratory (AFRL) where he was a program manager and principle scientist for the computational intelligence and neuromorphic computing research efforts. He also worked at IBM as an advisory scientist/engineer development enabling advanced CMOS technologies and as a business analyst within IBM's photomask business unit. Dr. Pino also served as an adjunct professor at the University of Vermont where he taught electrical engineering courses.

Dr. Pino has a B.E. in Electrical Engineering from the City University of New York and an M.Sc. and a Ph.D. in Electrical Engineering from the Rensselaer Polytechnic Institute. He is the recipient of numerous awards and professional distinctions; has published more than 40 technical papers, including three books; and holds six patents, three pending.

This work is dedicated to Dr. Pino's loving and supporting wife without whom this work would not be possible.

ICF International, Fairfax, USA   Dr. Robinson E. Pino

Contents

Towards Fundamental Science of Cyber Security 1
Alexander Kott

Bridging the Semantic Gap: Human Factors in Anomaly-Based
Intrusion Detection Systems 15
Richard Harang

Recognizing Unexplained Behavior in Network Traffic 39
Massimiliano Albanese, Robert F. Erbacher, Sushil Jajodia,
C. Molinaro, Fabio Persia, Antonio Picariello, Giancarlo Sperlì
and V. S. Subrahmanian

Applying Cognitive Memory to CyberSecurity 63
Bruce McCormick

Understanding Cyber Warfare 75
Yan M. Yufik

Design of Neuromorphic Architectures with Memristors 93
Dhireesha Kudithipudi, Cory Merkel, Mike Soltiz, Garrett S. Rose
and Robinson E. Pino

Nanoelectronics and Hardware Security 105
Garrett S. Rose, Dhireesha Kudithipudi, Ganesh Khedkar,
Nathan McDonald, Bryant Wysocki and Lok-Kwong Yan

User Classification and Authentication for Mobile Device Based
on Gesture Recognition 125
Kent W. Nixon, Yiran Chen, Zhi-Hong Mao and Kang Li

**Hardware-Based Computational Intelligence for Size, Weight,
and Power Constrained Environments**   137
Bryant Wysocki, Nathan McDonald, Clare Thiem, Garrett Rose
and Mario Gomez II

Machine Learning Applied to Cyber Operations   155
Misty Blowers and Jonathan Williams

Detecting Kernel Control-Flow Modifying Rootkits   177
Xueyang Wang and Ramesh Karri

**Formation of Artificial and Natural Intelligence in Big
Data Environment**   189
Simon Berkovich

**Alert Data Aggregation and Transmission Prioritization
over Mobile Networks**   205
Hasan Cam, Pierre A. Mouallem and Robinson E. Pino

Semantic Features from Web-Traffic Streams   221
Steve Hutchinson

Concurrent Learning Algorithm and the Importance Map   239
M. R. McLean

Hardware Accelerated Mining of Domain Knowledge   251
Tanvir Atahary, Scott Douglass and Tarek M. Taha

Memristors and the Future of Cyber Security Hardware   273
Michael J. Shevenell, Justin L. Shumaker and Robinson E. Pino

Towards Fundamental Science of Cyber Security

Alexander Kott

1 Introduction

Few things are more suspect than a claim of the birth of a new science. Yet, in the last few years, terms like "Science of Cyber," or "Science of Cyber Security," or "Cyber Science" have been appearing in use with growing frequency. For example, the US Department of Defense defined "Cyber Science" as a high priority for its science and technology investments [1], and the National Security Agency has been exploring the nature of the "science of cybersecurity" in its publications, e.g., [2]. This interest in science of cyber is motivated by the recognition that development of cyber technologies is handicapped by the lack of scientific understanding of the cyber phenomena, particularly the fundamental laws, theories, and theoretically-grounded and empirically validated models [3]. Lack of such fundamental knowledge—and its importance—has been highlighted by the US President's National Science and Technology Council [4] using the term "cybersecurity science."

Still, even for those in the cyber security community who agree with the need for science of cyber—whether it merits an exalted title of a new science or should be seen merely as a distinct field of research within one or more of established sciences—the exact nature of the new science, its scope and boundaries remain rather unclear.

This chapter offers an approach to describing this scope in a semi-formal fashion, with special attention to identifying and characterizing the classes of problems that the science of cyber should address. In effect, we will map out the landscape of the science of cyber as a coherent classification of its characteristic problems. Examples of current research—mainly taken from the portfolio of the United States Army Research Laboratory where the author works—will illustrate selected classes of problems within this landscape.

A. Kott (✉)
US Army Research Laboratory, Adelphi, MD, USA
e-mail: alexander.kott1.civ@mail.mil

R. E. Pino (ed.), *Network Science and Cybersecurity*,
Advances in Information Security 55, DOI: 10.1007/978-1-4614-7597-2_1,
© Springer Science+Business Media New York 2014

2 Defining the Science of Cyber Security

A research field—whether or not we declare it a distinct new science—should be characterized from at least two perspectives. First is the domain or objects of study, i.e., the classes of entities and phenomena that are being studied in this research field. Second is the set of characteristic problems, the types of questions that are asked about the objects of study. Related examples of attempts to define a field of research include [5] and [6].

To define the domain of the science of cyber security, let's focus on the most salient artifact within cyber security—malicious software. This leads us to the following definition: the domain of science of cyber security is comprised of phenomena that involve malicious software (as well as legitimate software and protocols used maliciously) used to compel a computing device or a network of computing devices to perform actions desired by the perpetrator of malicious software (the attacker) and generally contrary to the intent (the policy) of the legitimate owner or operator (the defender) of the computing device(s). In other words, the objects of research in cyber security are:

- Attacker A along with the attacker's tools (especially malware) and techniques T_a
- Defender D along with the defender's defensive tools and techniques T_d, and operational assets, networks and systems N_d
- Policy P, a set of defender's assertions or requirements about what event should and should not happen. To simplify, we may focus on cyber incidents I: events that should not happen.

Note that this definition of relevant domain helps to answer common questions about the relations between cyber security and established fields like electronic warfare and cryptology. Neither electronic warfare nor cryptology focus on malware and processes pertaining to malware as the primary objects of study.

The second aspect of the definition is the types of questions that researchers ask about the objects of study. Given the objects of cyber security we proposed above, the primary questions revolve around the relations between T_a, T_d, N_d, and I (somewhat similar perspective is suggested in [7] and in [8]. A shorthand for the totality of such relations might be stated as

$$(I, T_d, N_d, T_a) = 0 \qquad (1)$$

This equation does not mean we expect to see a fundamental equation of this form. It is merely a shorthand that reflects our expectation that cyber incidents (i.e., violations of cyber security policy) depend on attributes, structures and dynamics of the network of computing devices under attack, and the tools and techniques of defenders and attackers.

Let us now summarize what we discussed so far in the following definition. The science of cyber security is the study of relations between attributes, structures and dynamics of: violations of cyber security policy; the network of computing devices under attack; the defenders' tools and techniques; and the attackers' tools and techniques where malicious software plays the central role.

A study of relations between properties of the study's objects finds its most tangible manifestation in models and theories. The central role of models in science is well recognized; it can be argued that a science is a collection of models [9], or that a scientific theory is a family of models or a generalized schema for models [10, 11]. From this perspective, we can restate our definition of science of cyber security as follows. The science of cyber security develops a coherent family of models of relations between attributes, structures and dynamics of: violations of cyber security policy; the network of computing devices under attack; the defenders' tools and techniques; and the attackers' tools and techniques where malicious software plays the central role. Such models

1. are expressed in an appropriate rigorous formalism;
2. explicitly specify assumptions, simplifications and constraints;
3. involve characteristics of threats, defensive mechanisms and the defended network;
4. are at least partly theoretically grounded;
5. yield experimentally testable predictions of characteristics of security violations.

There is a close correspondence between a class of problems and the models that help solve the problem. The ensuing sections of this chapter look at specific classes of problems of cyber security and the corresponding classes of models. We find that Eq. 1 provides a convenient basis for deriving an exhaustive set of such problems and models in a systematic fashion.

3 Development of Intrusion Detection Tools

Intrusion detection is one of the most common subjects of research literature generally recognized as falling into the realm of cyber security. Much of research in intrusion detection focuses on proposing novel algorithms and architectures of intrusion detection tools. A related topic is characterization of efficacy of such tools, e.g., the rate of detecting true intrusions or the false alert rate of a proposed tool or algorithm in comparison with prior art.

To generalize, the problem addressed by this literature is to find (also, to derive or synthesize) an algorithmic process, or technique, or architecture of defensive tool that detects certain types of malicious activities, with given assumptions (often implicit) about the nature of computing devices and network being attacked, about the defensive policies (e.g., a requirement for rapid and complete identification of intrusions or information exfiltration, with high probability of success), and about the general intent and approaches of the attacker. More formally, in this problem we seek to derive T_d from N_d, T_a, and I, i.e.,

$$N_d, T_a, I \rightarrow T_d \qquad (2)$$

Recall that T_d refers to a general description of defenders' tools and techniques, that may include an algorithmic process or rules of an intrusion detection tool, as well as architecture of an IDS or IPS, and attributes of an IDS such as its detection rate. In other words, Eq. 2 is shorthand for a broad class of problems. Also note that Eq. 2 is derived from Eq. 1 by focusing on one of the terms on the left hand side of Eq. 1.

To illustrate the breadth of issues included in this class of problems, let's consider an example—a research effort conducted at the US Army Research Laboratory that seeks architectures and approaches to detection of intrusions in a wireless mobile network [12]. In this research, we make an assumption that the intrusions are of a sophisticated nature and are unlikely to be detected by a signature-matching or anomaly-based algorithm. Instead, it requires a comprehensive analysis and correlation of information obtained from multiple devices operating on the network, performed by a comprehensive collection of diverse tools and by an insightful human analysis.

One architectural approach to meeting such requirements would comprise multiple software agents deployed on all or most of the computing devices of the wireless network; the agents would send their observations of the network traffic and of host-based activities to a central analysis facility; and the central analysis facility would perform a comprehensive processing and correlation of this information, with participation of a competent human analyst (Fig. 1).

Fig. 1 Local agents on the hosts of the mobile network collect and sample information about hosts-based and network events; this information is aggregated and transmitted to the operation center where comprehensive analysis and detection are performed. Adapted from Ge et al. [12], with permission

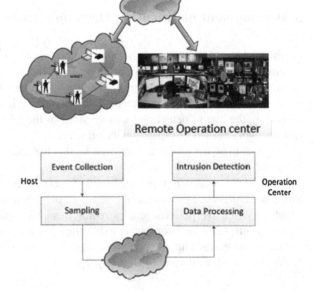

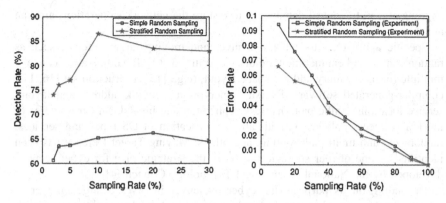

Fig. 2 Experiments suggest that detection rates and error rate of detection strongly depend on the traffic sampling ratio as well as the specific strategy of sampling. Adapted from Ge et al. [12], with permission

Such an approach raises a number of complex research issues. For example, because the bandwidth of the wireless network is limited by a number of factors, it is desirable to use appropriate sampling and in-network aggregation and pre-processing of information produced by the local software agents before trans-mitting all this information to the central facility. Techniques to determine appropriate locations for such intermediate aggregation and processing are needed. Also needed are algorithms for performing aggregation and pre-processing that minimize the likelihood of preserving the critical information indicating an intrusion. We also wish to have means to characterize the resulting detection accuracy in this bandwidth-restricted, mobile environment (Fig. 2).

Equation 2 captures key elements of this class of problems. For example, T_d in Eq. 2 is the abstraction of this defensive tool's structure (e.g., locations of interim processing points), behavior (e.g., algorithms for pre-processing), and attributes (e.g., detection rate). Designers of such a tool would benefit from a model that predicts the efficacy of the intrusion detection process as a function of architectural decisions, properties of the algorithms and properties of the anticipated attacker's tools and techniques.

4 Cyber Maneuver and Moving Target Defense

Cyber maneuver refers to the process of actively changing our network—its topology, allocation of functions and properties [13]. Such changes can be useful for several reasons. Continuous changes help to confuse the attacker and to reduce the attacker's ability to conduct effective reconnaissance of the network in prep-aration for an attack. This use of cyber maneuver is also called moving target defense. Other types of cyber maneuver could be used to minimize effects of an ongoing attack, to control damage, or to restore the network's operations after an attack.

Specific approaches to cyber maneuver and moving target defense, such as randomization and enumeration are discussed in [13, 14]. Randomization can take multiple forms: memory address space layout, (e.g., [15]; instruction set [16, 17]; compiler-generated software diversity; encryption; network address and layout; service locations, traffic patterns; task replication and breakdown across cores or machines; access policies; virtualization; obfuscation of OS types and services; randomized and multi-path routing, and others. Moving Target Defense has been identified as one of four strategic thrusts in the strategic plan for cyber security developed by the National Science and Technology Council [4].

Depending on its purpose, the cyber maneuver involves a large number of changes to the network executed by the network's defenders rapidly and potentially continuously over a long period of time. The defender's challenge is to plan this complex sequence of actions and to control its execution in such a way that the maneuver achieves its goals without destabilizing the network or confusing its users.

Until now, we used T_d to denote the totality of attributes, structure and dynamics of the defender's tools and techniques. Let's introduce additional notation, where ST_d is the structure of the defensive tools, and $BT_d(t)$ is the defender's actions. Then, referring to Eq. 1 and focusing on BT_d—the sub-element of T_d, the class of problems related to synthesis and control of defenders course of action can be described as

$$N_d, T_a, I \rightarrow BT_d(t) \tag{3}$$

An example of problem in this class is to design a technique of cyber maneuver in a mobile ad hoc spread-spectrum network where some of the nodes are compromised via a cyber attack and become adversary-controlled jammers of the network's communications. One approach is to execute a cyber maneuver using spread-spectrum keys as maneuver keys [18]. Such keys supplement the higher-level network cryptographic keys and provide the means to resist and respond to external and insider attacks. The approach also includes components for attack detection, identification of compromised nodes, and group rekeying that excludes compromised nodes (Fig. 3).

Equation 3 captures the key features of the problem: we wish to derive the plan of cyber maneuver $BT_d(t)$ from known or estimated changes in properties of our network, properties of anticipated or actually observed attacks T_a, and the objective of minimizing security violations I. Planning and execution of a cyber maneuver would benefit from models that predict relevant properties of the maneuver, such as its convergence to a desired end state, stability, or reduction of observability to the attacker.

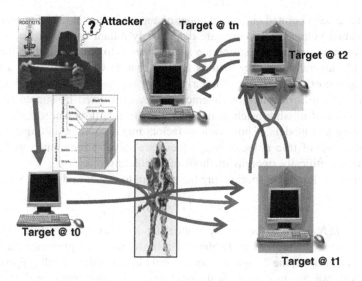

Fig. 3 In moving target defense, the network continually changes its attributes visible to the attacker, in order to minimize the attacker's opportunities for planning and executing an effective attack

5 Assessment of Network's Vulnerabilities and Risks

Monitoring and assessment of vulnerabilities and risks is an important part of cyber security strategy pursued by the US Government [19] This involves continuous collection of data through automated feeds including network traffic information as well as host information from host-based agents: vulnerability information and patch status about hosts on the network; scan results from tools like Nessus; TCP netflow data; DNS trees, etc. These data undergo automated analysis in order to assess the risks. The assessment may include flagging especially egregious vulnerabilities and exposures, or computing metrics that provide an overall characterization of the network's risk level. In current practice, risk metrics are often simple sums or counts of vulnerabilities and missing patches.

There are important benefits in automated quantification of risk, i.e., of assigning risk scores or other numerical measures to the network as a whole, its subsets and even individual assets [20]. This opens doors to true risk management decision-making, potentially highly rigorous and insightful. Employees at multiple levels—from senior leaders to system administrators—will be aware of continually updated risk distribution over the network components, and will use this awareness to prioritize application of resources to most effective remedial actions. Quantification of risks can also contribute to rapid, automated or semi-automated implementation of remediation plans.

However, existing risk scoring algorithms remain limited to ad hoc heuristics such as simple sums of vulnerability scores or counts of things like missing patches or open ports, etc. Weaknesses and potentially misleading nature of such metrics have been pointed out by a number of authors, e.g., [21, 22]. For example, the individual vulnerability scores are dangerously reliant on subjective, human, qualitative input, potentially inaccurate and expensive to obtain. Further, the total number of vulnerabilities may matters far less than how vulnerabilities are distributed over hosts, or over time. Similarly, neither topology of the network nor the roles and dynamics of inter-host interactions are considered by simple sums of vulnerabilities or missing patches. In general, there is a pronounced lack of rigorous theory and models of how various factors might combine into quantitative characterization of true risks, although there are initial efforts, such as [23] to formulate scientifically rigorous methods of calculating risks.

Returning to Eq. 1 and specializing the problem to one of finding N_d, we obtain

$$I, T_d, T_a \rightarrow N_d \qquad (4)$$

Recall that N_d refers to the totality of the defender's network structure, behavior and properties. Therefore, Eq. 3 refers to a broad range of problems including those of synthesizing the design, the operational plans and the overall properties of the network we are to defend. Vulnerabilities, risk, robustness, resiliency and controllability of a network are all examples of the network's properties, and Eq. 3 captures the problem of modeling and computing such properties.

An example of research on the problem of developing models of properties of robustness, resilience, network control effectiveness, and collaboration in networks is [24]. The author explores approaches to characterizing the relative criticality of cyber assets by taking into account risk assessment (e.g., threats, vulnerabilities), multiple attributes (e.g., resilience, control, and influence), network connectivity and controllability among collaborative cyber assets in networks. In particular, the interactions between nodes of the network must be considered in assessing how vulnerable they are and what mutual defense mechanisms are available (Fig. 4).

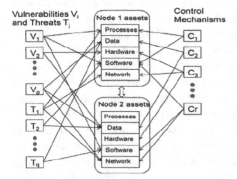

Contol Mechanisms	Weight	Node 1 assets	Node 2 assets	Node 3 assets
Signature-based NIDS	0.7	0	1	0
Host-based IDS	0.8	1	1	1
Anomaly-based IDS	0.6	1	1	0
Scanner	0.4	0	1	1
Firewall	0.6	0	0	1
Encryption	0.7	0	1	1
weighted asset value:		1.4	3.2	2.5

Fig. 4 Risk assessment of a network must take into account complex interaction between nodes of the network, particularly the interactions between their vulnerabilities as well as opportunities for mutual defense. Adapted from [24], with permission

6 Attack Detection and Prediction

Detection of malicious activities on networks is among the oldest and most common problems in cyber security [25]. A broad subset of such problems is often called intrusion detection. Approaches to intrusion detection are usually divided into two classes, signature-based approaches and anomaly-based approach, both with their significant challenges [26, 27]. In Eq. 1, the term I refers to malicious activities or intrusions, including structures, behaviors and properties of such activities. Therefore, the process of determining whether a malicious activity is present and its timing, location and characteristics, are reflected in the following expression:

$$T_d, N_d, T_a \rightarrow I \tag{5}$$

The broad class of problems captured by Eq. 5 includes the problem of deriving key properties of a malicious activity, including the very fact of an existence of such an activity, from the available information about the tools and techniques of the attacker T_a (e.g., the estimated degree of sophistication and the nature of past attempted attacks of the likely threats), tools and techniques of the defender T_d (e.g., locations and capabilities of the firewalls and intrusion-prevention systems), and the observed events on the defender's network N_d (e.g., the alerts received from host based agents or network based intrusion detection systems).

Among the formidable challenges of the detection problem is the fact that human analysts and their cognitive processes are critical components within the modern practices of intrusion detection. However, the human factors and their properties in cyber security have been inadequately studied and are poorly understood [28, 29].

Unlike the detection problem that focuses on identifying and characterizing malicious activities that have already happened or at least have been initiated, i.e., $I(t)$ for $t < t_{now}$, the prediction problem seeks to characterize malicious activities that are to occur in the future, i.e., $I(t)$ for $t > t_{now}$. The extent of research efforts and the resulting progress has been far less substantial in prediction than in detection. Theoretically grounded models that predict characteristics of malicious activities I—including the property of detectability of the activity—as a function of T_a, T_d, N_a, would be major contributors into advancing this area of research.

An example of research on identifying and characterizing probable malicious activities, with a predictive element as well, is [30], where the focus is on fraudulent use of security tokens for unauthorized access to network resources. Authors explore approaches to detecting such fraudulent access instances through a network-based intrusion detection system that uses a parsimonious set of information. Specifically, they present an anomaly detection system based upon IP addresses, a mapping of geographic location as inferred from IP address, and usage timestamps. The anomaly detector is capable of identifying fraudulent token usage with as little as a single instance of fraudulent usage while overcoming the often significant limitations in geographic IP address mappings. This research finds

Fig. 5 There exist multiple complex patterns of time-distance pairs of legitimate user's subsequent log-ins. The figure depicts the pattern of single-location users combined with typical commuters that log-in from more than one location. Adapted from Harang and Glodek [30], with permission

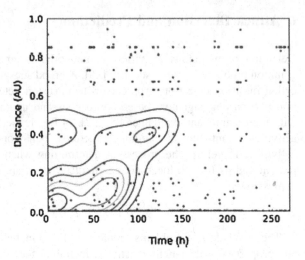

significant advantages in a novel unsupervised learning approach to authenticating fraudulent access attempts via time/distance clustering on sparse data (Fig. 5).

7 Threat Analysis and Cyber Wargaming

Returning once more to Eq. 1, consider the class of problems where the tools and techniques of the attacker Ta are of primary interest:

$$I, T_d, N_d, \rightarrow T_a \tag{6}$$

Within this class of problems we see for example the problem of deriving structure, behavior and properties of malware from the examples of the malicious code or from partial observations of its malicious activities. Reverse engineering and malware analysis, including methods of detecting malware by observing a code's structure and characteristics, fall into this class of problems.

A special subclass of problems occurs when we focus on anticipating the behavior of attacker over time as a function of defender's behavior:

$$I(t), T_d(t), N_d, \rightarrow T_a(t) \tag{6a}$$

In this problem, game considerations are important—both the defender's and the attacker's actions are at least partially strategic and depend on their assumptions and anticipations of each other's actions. Topics like adversarial analysis and reasoning, wargaming, anticipation of threat actions, and course of action development fall into this subclass of problems.

8 Summary of the Cyber Science Problem Landscape

We now summarize the classification of major problem groups in cyber security. All of these derive from Eq. 1. For each subclass, an example of a common problem in cyber security research and practice is added, for the sake of illustration.

$T_d, T_a, I \rightarrow N_d$

　　$T_d, T_a, I \rightarrow SN_d(t)$—e.g., synthesis of network's structure
　　$T_d, T_a, I \rightarrow BN_d(t)$—e.g., planning and anticipation of network's behavior
　　$T_d, T_a, I \rightarrow PN_d(t)$—e.g., assessing and anticipating network's security properties

$N_d, T_a, I \rightarrow T_d$

　　$N_d, T_a, I \rightarrow ST_d(t)$—e.g., design of defensive tools, algorithms
　　$N_d, T_a, I \rightarrow BT_d(t)$—e.g., planning and control of defender's course of action
　　$N_d, T_a, I \rightarrow PT_d(t)$—e.g., assessing and anticipating the efficacy of defense

$T_d, N_d, I \rightarrow T_a$

　　$T_d, N_d, I \rightarrow ST_a(t)$—e.g., identification of structure of attacker's code or infrastructure
　　$T_d, N_d, I \rightarrow BT_a(t)$—e.g., discovery, anticipation and wargaming of attacker' actions
　　$T_d, N_d, I \rightarrow PT_a(t)$—e.g., anticipating the efficacy of attacker's actions

$T_d, N_d, T_a \rightarrow I$

　　$T_d, N_d, T_a \rightarrow I(t), t < t_{now}$—e.g., detection of intrusions that have occured
　　$T_d, N_d, T_a \rightarrow I(t), t > t_{now}$—e.g., anticipation of intrusions that will occur

9 Conclusions

As a research filed, the emerging science of cyber security can be defined as the search for a coherent family of models of relations between attributes, structures and dynamics of: violations of cyber security policy; the network of computing devices under attack; the defenders' tools and techniques; and the attackers' tools and techniques where malicious software plays the central role. As cyber science matures, it will see emergence of models that should: (a) be expressed in an appropriate rigorous formalism; (b) explicitly specify assumptions, simplifications and constraints; (c) involve characteristics of threats, defensive mechanisms and the defended network; (c) be at least partly theoretically grounded; and (d) yield experimentally testable predictions of characteristics of security violations. Such models are motivated by key problems in cyber security. We propose and systematically derive a classification of key problems in cyber security, and illustrate with examples of current research.

References

1. Z.J. Lemnios, Testimony before the United States house of representatives committee on armed services, Subcommittee on Emerging Threats and Capabilities (2011), http://www.acq.osd.mil/chieftechnologist/publications/docs/ASDRE_Testimony_2011.pdf. Accessed 1 Mar 2011
2. T. Longstaff, Barriers to achieving a science of cybersecurity, Next Wave **19**(4), (2012), http://www.nsa.gov/research/tnw/tnw194/article5.shtml
3. JASON, The science of cyber security. Report JSR-10-102, MITRE, McLean VA (2010), http://www.fas.org/irp/agency/dod/jason/cyber.pdf
4. National Science and Technology Council, Trustworthy cyberspace: strategic plan for the federal cybersecurity research and development program (2011), http://www.nitrd.gov/fileupload/files/Fed_Cybersecurity_RD_Strategic_Plan_2011.pdf
5. J. Willis, Defining a field: content, theory, and research issues. Contemporary Issues in Technology and Teacher Education [Online serial] **1**(1) (2000), http://www.citejournal.org/vol1/iss1/seminal/article1.htm
6. H. Bostrom, et al., On the definition of information fusion as a field of research, in Technical Report (University of Skovde, School of Humanities and Informatics, Skovde, Sweden 2007), http://www.his.se/PageFiles/18815/Information%20Fusion%20Definition.pdf
7. F.B. Schneider, Blueprint for a science of cybersecurity. Next Wave **19**(2), 27–57 (2012)
8. J. Bau, J.C. Mitchell, Security modeling and analysis. Secur. Priv. IEEE **9**(3), 18–25 (2011)
9. R. Frigg, Models in science, Stanford Encyclopedia of Philosophy, 2012, http://plato.stanford.edu/entries/models-science/
10. Nancy Cartwright, *How the Laws of Physics Lie* (Oxford University Press, Oxford, 1983)
11. Patrick Suppes, *Representation and Invariance of Scientific Structures* (CSLI Publications, Stanford, 2002)
12. L. Ge, H. Liu, D. Zhang; W. Yu, R. Hardy, R. Reschly, On effective sampling techniques for host-based intrusion detection in *MANET, Military Communications Conference – MILCOM 2012* (2012)
13. S. Jajodia, A.K. Ghosh, V. Swarup, C. Wang, X.S. Wang (eds.), *Moving Target Defense: Creating Asymmetric Uncertainty for Cyber Threats*, Advances in Information Security, vol. 54 (Springer, Berlin, 2011)
14. S. Jajodia, A.K. Ghosh, V.S. Subrahmanian, V. Swarup, C. Wang, X.S. Wang (eds.), *Moving Target Defense: Application of Game Theory & Adversarial Modeling*, Advances in Information Security, vol. 100 (Springer, Berlin, 2013)
15. H. Bojinov et al., Address space randomization for mobile devices, in *Proceedings of Fourth ACM Conference on Wireless Network Security*, 2011, pp. 127–138
16. E.G. Barrantes et al., Randomized instruction set emulation. ACM Trans. Inf. Syst. Secur. **8**(1), 3–30 (2005)
17. S. Boyd, G. Kc, M. Locasto, A. Keromytis, V. Prevelakis, On the general applicability of instruction-set randomization'. IEEE Trans. Dependable Secure Comput. **7**(3), 255–270 (2010)
18. D. Torrieri, S. Zhu, S. Jajodia, *Cyber Maneuver Against External Adversaries and Compromised Nodes*, Moving Target Defense – Advances in Information Security, vol. 100 (Springer, New York, 2013), pp. 87–96

19. K. Dempsey, et al., *Information Security Continuous Monitoring ISCM_ for Federal Information Systems and Organizations* (NIST Special Publication, Gaithersburg, MD, 2011), pp. 800–137

20. A. Kott, C. Arnold, Promises and challenges of continuous monitoring and risk scoring. IEEE Priv. Secur. **11**(1), 90–93 (2013)

21. W. Jensen, Directions in Security Metrics Research, National Institute of Standards and Technology, (NISTIR 7564), Apr 2009

22. N. Bartol et al., Measuring cyber security and information assurance: a state of the art report, Defense Technical Information Center, May 2009

23. R.P. Lippman, et al., Continuous security metrics for prevalent network threats: introduction and first four metrics, Technical Report ESCTR- 2010-090, MIT, May 2012

24. H. Cam, PeerShield: determining control and resilience criticality of collaborative cyber assets in networks, in *Proceedings of SPIE 8408, Cyber Sensing 2012, 840808* (1 May 2012)

25. J.P. Anderson, *Computer Security Threat Monitoring and Surveillance* (James P. Anderson Co., Fort Washington, 1980)

26. Stefan Axelsson, The base-rate fallacy and the difficulty of intrusion detection. ACM Trans. Inf. Syst. Secur. **3**(3), 186–205 (2000)

27. Animesh Patcha, Jung-Min Park, An overview of anomaly detection techniques: existing solutions and latest technological trends. Comput. Netw. **51**(12), 3448–3470 (2007)

28. M. McNeese, Perspectives on the role of cognition in cyber security, in *Proceedings of the Human Factors and Ergonomics Society 56th Annual Meeting*, vol. 56, 2012, p. 268

29. M. Boyce, K. Duma, L. Hettinger, T. Malone, D. Wilson, J. Lockett-Reynolds, Human performance in cyber security: a research agenda. in *Proceedings of the Human Factors and Ergonomics Society 55th Annual Meeting*, vol. 55, 2011, p. 1115

30. R.E. Harang, W.J. Glodek, Identification of anomalous network security token usage via clustering and density estimation, in *46th Annual Conference on Information Sciences and Systems (CISS)*, 21–23 Mar 2012, pp.1–6

Bridging the Semantic Gap: Human Factors in Anomaly-Based Intrusion Detection Systems

Richard Harang

1 Introduction

Network intrusion detection systems (IDSs) are often broadly divided into two broad groups; those that attempt to detect "misuse" (typically including signature-based approaches), and those that attempt to detect "anomalies" [1, 2]. Misuse detection typically relies on static rules (a.k.a. "signatures") to detect malicious activity [3, 4]; activity that precisely matches a clearly defined set of rules is flagged as potentially malicious and referred for further action/analysis, while any activity that does not match any signature is assumed safe. This approach has so far proven quite effective, and the misuse detector Snort is perhaps one of the most widely deployed IDSs in use today. The misuse detection approach generally (although not always, see, e.g., [5]) benefits from an acceptably low false positive rate, and assuming adequate computational resources, is capable of detecting any attack it has a signature for. This low false positive rate comes at a price, however; due to the highly specific nature of the signatures that misuse detection requires, it is typically ineffective in detecting novel "zero-day" attacks targeting new vulnerabilities, for which no effective signature has yet been developed [6] and requires constant maintenance of the signature database obfuscated variants are developed (although see [7] for convincing arguments that anomaly-based detection will likely fare no better).

Anomaly-based detection generally attempts to fill a complementary role by applying the powerful and (in other domains) highly successful tools of machine learning and statistical analysis to the intrusion detection problem in order to detect variants on existing attacks or entirely new classes of attacks [8]. Such methods have become fairly widespread in the IDS academic literature, however very few anomaly detection systems appear to have been deployed to actual use [1], and none appear to be adopted on anything near the scale of misuse systems such as Snort [3] or Bro [4]. While a brief literature review will reveal hundred of papers examining methods

R. Harang (✉)
ICF International, Washington, DC, USA
e-mail: Richard.Harang@ICFI.com

R. E. Pino (ed.), *Network Science and Cybersecurity*,
Advances in Information Security 55, DOI: 10.1007/978-1-4614-7597-2_2,
© Springer Science+Business Media New York 2014

ranging from self-organizing maps [9] to random forests [10] to active learning methods [11], there are very few that discuss operational deployment or long-term use of any of these approaches. Several potential reasons for this disparity are presented in [1], including their unacceptably high false positive rates (see also discussion in [12]), the severely skewed nature of the classification problem that IDS research presents, the lack of useful testing data, and the regrettable tendency of researchers in anomaly detection to treat the IDS as a black box with little regard to its operational role in the overall IDS process or the site-specific nature of security policies (referred to by [1] as the "semantic gap").

In the following, we begin by examining the techniques used in several anomaly IDS implementations in the literature, and provide some discussion of them in terms of the "semantic gap" of Somners and Paxson. We then provide a sketch of the base rate fallacy argument as it applies to intrusion detection, originally presented by Axelsson in [12]. We then extend the results of [12] briefly to show that the utility of an anomaly IDS is in fact closely related to the semantic gap as well as the false positive rate. Finally, we return to tree-based classifiers, present anecdotal evidence regarding their interpretability, and discuss methods for extracting rules from randomly constructed ensembles that appear to permit anomaly detection methods to greatly facilitate human interpretation of their results at only modest costs in accuracy.

2 Anomaly IDS Implementations

The field of anomaly detection has seen rapid growth in the past two decades, and a complete and exhaustive inventory of all techniques is prohibitive. We attempt to summarize the most popular implementations and approaches here with illustrative examples, rather than provide a comprehensive review.

Anomaly detection itself may be split into the two general categories of supervised and unsupervised learning [2, 8]. Supervised learning remains close to misuse detection, using a "training set" of known malicious and benign traffic in order to identify key predictors of malicious activity that will allow the detector to learn more general patterns that may detect both previously seen attacks as well as novel attacks not represented in the original training set. Unsupervised learning often relies on outlier detection approaches, reasoning that most use of a given system should be legitimate, and data points that appear to stand out from the rest of the data are more likely to represent potentially hostile activities.

2.1 Categorical Outlier Detection

In [13], a fairly general method for outlier detection in tabular categorical data is presented in which—for each column in each row—the relative frequency of the entry is compared to all other entries in the given column; this method does not consider covariance between columns, and so has obvious links to e.g., Naïve Bayes models, but can be operated in $O(nm)$ time where n is the number of rows (data points) in the table, and m the number of columns (attributes). This method is most appropriate for purely categorical data where no metric or ordering exists, however they demonstrate that discretization of numerical data allows application of their method.

In [14], a slightly more general method is presented to allow for outlier detection on mixed categorical and numerical data, while permitting a varying number of categorical attributes per item, by examining set overlap for categorical attributes while monitoring conditional covariance relationships between the numerical attributes and performing weighted distance calculations between them. They also apply their method to the KDD'99 intrusion detection data set to examine its performance, reporting false positive rates of 3.5 %, with widely varying true positive rates per class of attack (from 95 to 0 %).

In the strictly continuous domain, the work of [15] examines conditional anomaly detection, i.e., detecting anomalous entries based on their context, by means of estimation from strictly parametric models. They fit a Gaussian mixture model to their observed data via maximum likelihood using the expectation–maximization algorithm, and detect outliers on the basis of low likelihood score. The work of [15] draws a rather clear distinction between "environment" variables that are conditioned on, and "indicator" variables that are examined for anomalies, however this might be readily extended by regenerating their models in sequence, once for each variable.

This class of outlier detection method generally forms one of the more inter-pretable ones; because individual points are generally considered in their original coordinate system (if one exists, c.f. support vector machines, below) or in terms of marginal probabilities, the explanation for why a particular point was classified as anomalous is generally amenable to introspection. For instance, in the work of [13], each field of an 'anomalous' entry may be inspected for its contribution to the score, and the ones contributing most to the score may be readily identified. The work of [14] is slightly more complex, due to the individualized covariance scoring, however the notion of set overlap is straightforward and the covariance relationship between variables lends itself to straightforward explanation. Unfor-tunately, despite their explanatory power, these methods generally do not appear to provide comparable performance to more complex methods such as those described below.

2.2 Support Vector Machines

Both supervised and unsupervised methods using support vector machines (SVMs) [16] have become extremely widespread due to the good generalization perfor-mance and high flexibility they afford. One-class SVMs are used in [17] to detect "masquerades", illicit use of a legitimate account by a third party posing as the legitimate user, and found to compare favorably to simpler techniques such as Naïve Bayes. One-class SVMs are also used in conjunction with conventional signature based approaches in [18] to provide enhanced detection of variations on existing signatures.

Supervised learning with two-class SVMs has also been applied to the KDD'99 data set in [19], and found to compare favorably to neural network techniques with respect to training time, however as the most widely used SVMs are by construction binary classifiers (multi-class SVMs are typically constructed on a 1-vs-all basis from multiple binary SVMs; SVM constructions that explicitly allow multiple classes exist but are computationally expensive [20]).

SVMs are also commonly used as a component of more complex systems; these hybrid approaches allow for the generalization power of SVMs to be exploited, while mitigating some of their common weaknesses, such as poor performance when dealing with ancillary statistics. In [21], the authors employ rough sets to reduce the feature space presented to SVMs for training and testing. A feature extraction step is also used to reduce the input space to one-class SVMs in [6], allowing them to better control the false positive rate.[1] Finally, [22] use a genetic algorithm to do initial feature extraction for a SVM, also using KDD'99 data; they also attempted to apply their method to real data, however concluded that their own network data contained too few attacks to reliably evaluate their method.

While strictly speaking, SVMs are simply maximum margin linear classifiers, SVMs as commonly used and understood derive their excellent classification properties from the "Kernel trick", which uses the fact that data can be projected into certain inner product spaces (reproducing kernel Hilbert spaces) in which inner products may be computed through a kernel function on the original data without having to project the data into the space. These spaces are typically significantly more complex than the original space, but the kernel trick allows any classification method that may be written entirely in terms of inner products (such as support vectors) to be applied to those complex spaces based entirely on the original data, thus finding a separating hyperplane in the transformed space without ever explicitly constructing it.

While this projection into one of a wide range of high dimensional spaces gives SVMs their power, it also poses significant barriers to interpretability. Visualization and interpretation of (for instance) a 41-dimensional separating hyperplane is well beyond the capabilities of most people; developing an intuition for the effects of projection into what may be in principle an infinite dimensional space, and then considering the construction of a separating hyperplane in this space complicates matters substantially. For illustrative purposes, we have combined the approach of [23] (see below) with one-class support vector machines, and applied

---

[1] It is also worth noting that [6] use and provide access to a "KDD-like" set of data—that is, data aggregated on flow-like structures containing labels—that contains real data gathered from honeypots between 2006 and 2009 rather than synthetic attacks that predate 1999; this may provide a more useful and realistic alternative to the KDD'99 set.

```
###[ IP ]###
          version   = 4L
          ihl       = 5L
          tos       = 0x8
          len       = 1500
          id        = 47564
          flags     = DF
          frag      = 0L
          ttl       = 63
          proto     = tcp
          chksum    = 0x921d
          src       = 10.2.20.40
          dst       = 10.2.193.254
###[ TCP ]###
             sport     = 52711
             dport     = neod1
             seq       = 137580262
             ack       = 1222080868
             dataofs   = 5L
             reserved  = 0L
             flags     = A
             window    = 5840
             chksum    = 0xb0c1
             urgptr    = 0
             options   = []
###[ Raw ]###
0000   34 6E DD E9 76 BB 9F CD  19 8E 2A 46 95 87 72 38
0010   AD 47 BD 61 0C F5 B4 26  38 C7 6B 31 6B F5 6C FC
0020   98 2B 98 39 F1 56 61 F7  19 D5 B1 06 58 B4 D7 C4
0030   B8 DD D9 F0 0E 81 82 B1  7C 3A DE 53 EE 80 BB 27
0040   8E 73 2E 81 C5 C4 1D 22  E3 CE 07 5F D8 51 CF 88
...et cetera
```

Fig. 1 A packet within the support of the 1-class SVM

it to packet capture data from the West Point/NSA Cyber Defense Exercise [24]. Briefly, the histogram of 3-byte n-grams from each packet was constructed as a feature vector and provided to the 1-class SVM function within Scikit-Learn [25]; default values were used, and built-in functions were used to recover on support vector data point (Fig. 3). The norm imposed by the inner product derived from the kernel function was used to find the nearest point contained within the support of the SVM (Fig. 1), as well as the nearest point outside the support of the SVM (Fig. 2).[2] These figures provide Scapy [26] dissections of the three packets.

While the mathematical reason that the packet in Fig. 1 is considered 'normal' while that of Fig. 2 would be considered 'anomalous' is straightforward, from the

---

[2] A possibly instructive exercise for the reader: obscure the labels and ask a knowledgeable colleague to attempt to divine which two packets are 'normal' and which is 'anomalous', and why.

```
###[ IP ]###
        version   = 4L
        ihl       = 5L
        tos       = 0x8
        len       = 1500
        id        = 47565
        flags     = DF
        frag      = 0L
        ttl       = 63
        proto     = tcp
        chksum    = 0x921c
        src       = 10.2.20.40
        dst       = 10.2.193.254
###[ TCP ]###
           sport    = 52711
           dport    = neod1
           seq      = 137581722
           ack      = 1222080868
           dataofs  = 5L
           reserved = 0L
           flags    = A
           window   = 5840
           chksum   = 0xcc4a
           urgptr   = 0
           options  = []
###[ Raw ]###
0000   C0 1C 65 35 A2 B9 44 C3   E0 30 40 BC E8 11 23 66
0010   A8 FF 6C 98 B6 82 10 C8   3A 99 DA 1E 25 44 B3 10
0020   62 07 F8 3C DD 15 FB 50   84 A2 04 20 F7 9B 0F E5
0030   C6 FA 61 EE FB 07 4B 22   0E 0A 6A 4E 24 B1 F1 2A
0040   6D 8F 85 80 E5 C3 6B 03   19 62 C2 B9 59 CD 45 91
...and so on
```

Fig. 2 A packet outside the support of the 1-class SVM

point of view of a human analyst, dealing with this output is difficult. In particular the transformed space of the SVM does not map neatly to the normal space of parameters that human analysts work within to determine hostile intent in network traffic, and the region of high-dimensional space, even assuming that analysts are willing and able to visualize point clouds in infinite dimensional space, does not neatly map to threat classes, misconfigurations, or other common causes of undesired activity. In effect, to either confirm or refute the classification of this anomaly detector, analysts must launch an entirely independent investigation into the packet at issue, with very little useful information on where the most fruitful areas for examination are likely to be.

```
###[ IP ]###
          version   = 4L
          ihl       = 5L
          tos       = 0x8
          len       = 1500
          id        = 28176
          flags     = DF
          frag      = 0L
          ttl       = 63
          proto     = tcp
          chksum    = 0xddd9
          src       = 10.2.20.40
          dst       = 10.2.193.254
   ###[ TCP ]###
             sport    = 32900
             dport    = afrog
             seq      = 83970169
             ack      = 2996538417
             dataofs  = 5L
             reserved = 0L
             flags    = A
             window   = 5840
             chksum   = 0x3a3b
             urgptr   = 0
             options  = []
   ###[ Raw ]###
      0000   5C 73 37 78 61 81 9F 17   27 29 A2 58 6C 83 5A CC
      0010   F8 3B F5 47 8B 9A B3 81   DF 63 D7 82 84 9A AF 9A
      0020   24 9B 7B D7 2E 5A D4 00   2A 37 E0 56 45 B1 4D 65
      0030   7E 22 9E A5 F1 D8 6E 34   1B 8A 7C CF 36 E5 5F F4
      0040   AD 2D 17 64 16 10 1A 33   E5 C1 8A 39 29 FF 06 78
      0050   55 3C D8 4A 73 CB 28 E2   6D 8E 29 DA 2A 5C 08 B5
```

Fig. 3 The support vector

2.3 Clustering and Density Estimation

One of the more common forms of unsupervised learning is clustering. Such approaches in untransformed coordinates have also been examined; the work of [27] presents an application of an ensemble approach in which k-means clustering is used to on a per-port basis to form representative clusters for the data; new data is then assigned to a cluster which may accept or reject it as anomalous. In-place incremental updating of the clusters and their associated classification rules is used, and while the approach appears to underperform when compared to other, more complex ones applied to the KDD'99 data with a false positive rate of roughly 10 %, the method of cluster assignment and subsequent determination provides a degree of introspection to the decision not available to the more complex approaches.

The work of [23] addresses the problem of transforming the highly variable and often difficult to model content of individual network packets into a continuous space amenable to distance-based outlier detection via frequency counts of n-grams. For a given length n, all possible consecutive n-byte sequences are extracted from a packet and tabulated into a frequency table, which is then used to define a typical distribution for a given service (implicitly assuming a high-dimensional multivariate normal). Once this distribution has stabilized, further packets are then examined for consistency with this distribution, and ones that differ significantly are identified as anomalous. While the reported performance of the classifier is excellent (nearly 100 % accuracy on the DARPA IDS data set that formed the basis for the KDD'99 data), once packets are grouped per-port and per-service, the question of interpretability once again becomes a difficult one. While a priori known attacks in the data, including novel ones, were detected by their method, they do not discuss the issue of determining the true/false positive status of an alarm in unlabeled data in any detail.

2.4 Hybrid Techniques

The work of [11] represents an interesting case, which uses a combination of simulated data and active learning to convert the outlier detection problem into a more standard classification problem. Simulated "outlier" data is constructed either from a uniform distribution across a bounded subspace, or from the product distribution of the marginal distributions of the data, assuming independence, and adjoined to the original data with a label indicating its synthetic nature (see also [28], which uses random independent permutations of the columns to perform a similar task). The real and simulated data, labeled as such, is then passed to an ensemble classifier that attempts to maximize the margin between the two classes, using sampling to select points from regions where the margin is small to refine the separation between the classes. This method is also applied to KDD'99 data, where the results are actually shown to outperform the winning supervised method from the original contest. In discussing the motivation for their method, they point out that a "notable issue with [most outlier detection methods] … is the lack of explanation for outlier flagging decisions." While semantic issues are not addressed in further detail, the explicit classification-based approach they propose generates a reference distribution of 'outliers' that can be used as an aid to understand the performance of their outlier detection method.

Finally, the work of [9] combines both supervised and unsupervised learning approaches in order to attempt to leverage the advantages of each. They implement a C4.5 decision tree with labeled data to detect misuse, while using a self-organizing map (SOM) to detect anomalous traffic. Each new data point is presented to both techniques, and the output of both classifiers is considered in constructing the final classification of the point. While in this case the supervised learning portion of the detector could provide some level of semantic interpretation of anomalies, should both detectors fire simultaneously, the question of how to deal with alerts that are triggered only by the SOM portion of the detector is not addressed.

2.5 Tree-Based Classifiers

Tree-based ensemble classifiers present alternatives to kernel-based approaches that rely upon projections of the data into higher-dimensional spaces. From their initial unified presentation in [28] on through the present day [29], they have been shown to have excellent performance on a variety of tasks. While their original applications focused primarily on supervised learning, various methods to adapt them to unsupervised approaches such as density estimation and clustering have been developed since their introduction (the interested reader is referred to [29] and references therein for an overview of the variety of ways that random decision forests have been adapted to various problems, as well as an excellent selection of references). Unsurprisingly, random decision forests and variants thereof have been applied to the anomaly IDS problem as well.

A straightforward application of random decision forests in a supervised learning framework using the KDD'99 data is provided in [30], reporting a classification accuracy rate of 99.8 %. Similar work in [10] using feature selection rather than feature importance yields similar results, reporting classification accuracy of 99.9 %. Our own experiments have shown that simply blindly applying the built-in random decision forest package in Scikit-learn [25] to the first three features of the KDD'99 data in alphabetical order—"Count", "diff_srv_-rate", and "dst_bytes"—without any attempt to clean, balance, or otherwise adapt to the deficiencies of the KDD'99 set, yields over a 98 % accuracy rate in classifying KDD'99 traffic, with the bulk of the errors formed by misclassifications within the class of attacks (see Appendix for details).

An outlier detection approach for intrusion detection using a variant of random decision forests that the authors term "isolation forest" is presented in [31], in which the data is sub-sampled in extremely small batches (the authors recommend 256 points per isolation tree, and building a forest from 100 such trees), and iteratively split at random until the leaf size is reduced to some critical threshold. Using the intuition that anomalies should stand out from the data, it then follows that the fewer splits that are required to isolate a given point, the less similar to the rest of the data it is. Explicit anomaly scoring in terms of the expected path length of a binary search tree is computed. The authors of [31] examine—among other data sets—the portion of the KDD'99 data representing SMTP transactions, and report an AUC of 1.0 for the unsupervised detection of anomalies. Several of the same authors also present on-line versions of their algorithm in [32], taking advantage of the constant time complexity and extremely small memory footprint of their method to use sequential segments of data as training and test sets (where the test set from the previous iteration becomes the training set for the next iteration), and report similarly impressive results.

While the authors of [31] and [32] do not explicitly consider the notion of interpretation in their work, the earlier work of [33] provides an early example of an arguably similar approach that does explicitly consider such factors. Their anomaly detection method is explicitly categorical (with what amounts to a density estimation clustering approach to discretize continuous fields), and operates by generating a "rule forest" where frequently occurring patterns in commands and operations on host systems are used to generate several rule trees, which are then pruned based on quality of the rule and the number of observations represented by the rule.

Note, however, that in contrast to [31]—which also relies on trees—the approach of [33] lends more weight to longer rules that traverse more levels of the tree (where [31] searches for points that on average traverse very few levels of a tree before appearing in a leaf node). This occurs as the work of [31] (and outlier detection in general) tacitly assumes that the outlying data is not generated by the same process as the 'inlying' data, and hence the work of [31] searches for regions containing few points. Rule-based systems, such as [33] and [27] (and to a lesser extent, [34]) place more emphasis on conditional relationships in the data, in effect estimating the density for the final partitioning of the tree conditional on the prior path through the tree.

Also in contrast to many other anomaly IDS methods, [33] made interpretation of the results of their system—named "Wisdom and Sense"—an explicit design goal, noting that "anomaly resolution must be accomplished by a human." The rules themselves, represented by the conjunction of categorical variables, lend themselves much more naturally to human interpretation than e.g., the high-dimensional projections common to support vector machines, or the laborious transformation of categorical data into vectors of binary data which is then blindly projected into the corners of a high-dimensional unit cube.

3 Anomaly IDSs "in the wild" and the Semantic Gap

Likely due to the complexities enumerated above, experimental results on attempts to deploy anomaly detection systems are regrettably few and far between. Earlier methods, such as that of [33] or that of [34], typically report results only on their own in-house and usually proprietary data. When these systems have been tested elsewhere (see [35], in which [34] was tested and extended), many configuration details have been found to be highly site-specific, and additional tuning was required. As the volume and sensitivity of network traffic increased, along with the computational complexity of the anomaly detection methods used to construct anomaly detectors, attempts to deploy such detectors to operational environments do not appear to be widely reported. The work of [36] provides a notable exception to this rule, in which several proprietary anomaly detection systems were deployed in a backbone environment. Details about the anomaly detection methods used in the commercial products tested in [36] are unavailable (and some are noted as also incorporating 'signature based methods', though it is unclear at what level these signatures are being applied), however it is worth remarking that the bulk of the anomalies being examined in [36] were related to abnormal traffic patterns, rather than content (e.g., shell code, worms, Trojans, etc.). The volume of traffic and the extremely low rate of true positives meant that searching for more sophisticated attacks was essentially infeasible, and indeed the traffic patterns observed at the backbone level that were flagged as anomalous are much more straightforward to classify manually than more subtle exploitation attempts. Revisiting our terminology above, the cost α of evaluating a positive result and determining if it is a true or false positive is relatively low, at least in comparison to attacks that require content inspection to detect.

It is also worth remarking that in [36] it was observed that the set of true positive results on which the various anomaly detectors agreed was in fact extremely small, which suggests that the false negative rate for a single detector is in all likelihood rather high. This may a practical consequence of Axelsson's [12] result, suggesting that to retain a tractable false positive rate, the threshold for detection has been set rather high in these commercial anomaly detectors, thus increasing the false negative rate in tandem with the reduction in false positives, however in the absence of more details on the precise mode of operation and the numerical choices made in the algorithms, this remains speculative.

Issues with interpretability of anomaly IDS techniques are is explicitly brought up in [11], in which they point out that many outlier detection methods "...[tend] not to provide semantic explanation as to why a particular instance has been flagged as an outlier." The same observation within the context of IDS work has been made in some of the earliest work in anomaly-based IDS, including [33] (who in 1989 wrote "...explaining the meaning and likely cause of anomalous transactions ... must primarily be accomplished by a human.") and [35] (extending and testing the earlier work of [34]) who note that, in the absence of automated analysis of audit data, a "security officer (SO) must wade through stacks of printed output." They are also are discussed briefly in [1], where the example of [37] is given, in which the titular question "Why 6?" (or, more completely, why a subsequence length of 6 turns out to be the minimum possible value that permits the anomaly IDS to find the intrusions in the test set) ultimately requires 26 pages and a full publication to answer in a satisfying manner. While this does not directly address operational issues, the point remains that if the researchers who study and implement a system require that much effort to understand why a single parameter must be set the way it is to generate acceptable performance, it is not a question that analysts responsible for minute-to-minute monitoring and analysis of network traffic are likely to have the time or inclination to answer.

At least part of the blame for the resilience of the semantic gap may be laid at the feet of the availability of data. Analysis methods must be developed to fit the available data, and regrettably the most commonly-used data set in intrusion detection research remains the infamous KDD'99 set, which began receiving criticism as little as 4 years after its initial release (see [38] and [39], for instance) but—over the protests of many security researchers [24]—has been and continues to be widely used, particularly in the field of anomaly detection (see, e.g., [2, 9, 10, 22, 30, 32]).

The KDD'99 set is, superficially, extremely amenable to analysis by machine learning techniques (although see [38] and [39] for some notable issues); the data has 42 fields, the majority of which are continuously-valued, while the bulk of the remainder are Boolean. Only two fields, protocol type and service, are categorical, and both of these have a limited number of possible values (and, in practice, appear to be generally ignored in favor of continuous fields for most anomaly IDS research using this data). This enables and encourages an application of machine learning approaches that rely on metrics in inner product spaces (or their transformations) to anomaly IDS problems. The contextual information that human analysts often rely on to determine whether or not traffic is malicious (e.g., source and destination ports and IP blocks, protocol/port pairs, semantic irregularities such as malformed HTTP requests, and so forth) is very explicitly not included in this data, for the precise reason that it is very poorly suited to automated analysis by machine learning algorithms (recent work on language-theoretic security in fact suggests that automated analysis of many security problems is in fact *in principle* intractable, as it effectively reduces to solving the halting problem [40]).

The root of the semantic gap is thus in part the result of a vicious spiral: with the only widely used labeled data set being the KDD'99 data, which encourages the use of machine learning techniques that often rely on complex transformations of the data that tend to obscure any contextual information, the observation of [12] means that focus must be placed on reducing false positive rates to render them useful. This leads to more sophisticated and typically more complex methods, enlarging the semantic gap, leading to a greater emphasis on reduced false positive rate, and so forth.

4 The Base-Rate Fallacy, Anomaly Detection, and Cost of Misclassification

In addition to issues of interpretation, the simple fact that the overwhelming majority of traffic to most networks is not malicious has a significant impact on the reliability and usability of anomaly detectors. This phenomenon—the base-rate fallacy—and the impact it has on IDS systems is discussed in depth in [12]. We summarize selected relevant points of the argument here.

Briefly, the very low base rate for malicious behavior in a network leads to an unintuitive result, showing that the reliability of the detector (referred to in [12] as the "Bayesian detection rate") $P(I|A)$, or the probability that an incident of concern actually has occurred given the fact that an alarm has been produced, is overwhelmingly dominated by the false positive rate of a detector $P(A|\sim I)$, read as the probability that there is an alarm given that no incident of concern actually took place. A quick sketch by Bayes' theorem shows that:

$$P(I|A) = \frac{P(A|I)P(I)}{P(A|I)P(I) + P(A|\sim I)P(\sim I)}$$

where we assume all probabilities are with respect to some constant measure for a given site and IDS. If we then note that, by assumption, $P(I) \ll P(\sim I)$ for any given element under analysis, we can immediately see that the ratio on the right hand side is dominated by the $P(A|\sim I)P(\sim I)$ term. Since $P(\sim I)$ is assumed to be beyond our control, the only method of adjusting the reliability $P(I|A)$ is by adjusting the false positive rate of our detector.

This insight, has led to much work in recent results on reducing false positive rates in not just anomaly IDSs (see, e.g., [6] and [22]) but also in such misuse detectors as Snort [5, 18]. The availability of the KDD'99 set, which provides a convenient test-bed for such methods, despite its age and concerns about its reliability [38, 39], makes it relatively easy for research to focus on this issue, and so perhaps accounts for the popularity of KDD'99 despite its well-known issues.

The key insight from the base-rate fallacy argument is there is an extreme imbalance between the rate of occurrence of hostile traffic and that of malicious traffic, which in turn greatly exaggerates the impact of false positives on the reliability of the detector. However, if we extend the argument to include cost of classification, this same observation suggests a potential remediation. Assume that the cost of investigating any alarm, whether ultimately associated with an incident or not, is α, while the cost of not having an alarm when there is an incident (a false negative result) is β. We then have that

$$E[Cost] = \alpha P(A) + \beta P(\sim A|I)P(I)$$
$$= \alpha[P(A|I)P(I) + P(A|\sim I)P(\sim I)] + \beta P(\sim A|I)P(I)$$

And if we once again assume that $P(I) \ll P(\sim I)$, then note that we can approximate then term $\alpha[P(A|I)P(I) + P(A|\sim I)P(\sim I)]$ by simply $\alpha P(A|\sim I)P(\sim I)$, such that:

$$E[Cost] \approx \alpha P(A|\sim I)P(\sim I) + \beta P(\sim A|I)P(I)$$

We assume that the cost of false negatives—classifying hostile traffic as benign—may be large enough that we cannot ignore the second term. Note that the expected cost in this case is now very sensitive to even small changes in the cost of examining alarms. This suggests that in the event that the false positive rate cannot be reduced significantly, we may still be able to operate the system at an acceptable cost by reducing the cost of examining IDS alerts. As we discuss later sections, current anomaly detectors are not well-suited to this task, and in fact attempting to reduce the cost by reducing $P(A|\sim I)$ have in fact inadvertently led to increases in α, by virtue of both the need to adapt the heterogeneous data available to IDS systems to standard machine learning techniques that focus on inner product spaces, and the complex transforms of that data undertaken by these machine learning techniques in order to obtain accurate decision boundaries.

4.1 Cost Versus Threshold Tradeoffs

In many cases, the false positive rate and the false negative rate are related to each other; see, for instance, [6] where the effect of the support radius of a 1-class SVM on the true and false positive rate was examined. While shrinking the space mapping to a decision of 'anomaly' (or equivalently, increasing the detection threshold) clearly will reduce the false positive rate, it also decreases the sensitivity of the detector, and thus the true positive rate $P(A|I)$, and since $P(A|I) + P(\sim A|I) = 1$, i.e., every incoming packet must either trigger an alarm or not trigger an alarm, this must inevitably increase the false negative rate, potentially incurring a significant cost.

As a toy example, consider the Gaussian mixture problem, where our data comes from the following distribution:

$$f_X(x_i) = P(I)p(x_i; 1, 1) + (1 - P(I))p(x_i; 0, 1)$$

$$p(x_i; \mu, \sigma^2) = \frac{1}{\sqrt{2\pi\sigma^2}} \exp\left\{ -\frac{1}{2\sigma^2}(x_i - \mu)^2 \right\}$$

where for each x_i we must determine whether it was drawn from $p(x_i; 1, 1)$, in which case we deem it anomalous, or $p(x_i; 0, 1)$, in which case it is normal. Put another way, given some value x_i, our task is to decide whether $\mu_i = 1$ or $\mu_i = 0$. If we assume independence and construct a decision rule based solely on the current observation x_i, then we may fix a false negative rate of $P(\sim A|I) = p$; giving us a threshold for x_i of $x^\star = \Phi_{X|\mu=1}^{-1}(p)$ where Φ is the standard Gaussian CDF. From this we can obtain directly the false positive rate for this detector:

$$P(A|\sim I) = 1 - \Phi_{X|\mu_i=0}(x^\star)$$

Using this threshold-based decision rule, we have that the only possible way to decrease the false positive rate is to increase the threshold $x^\star$. As the CDF of x_i is by definition non-decreasing in x_i, we have immediately that increasing the threshold $x^\star$ will simultaneously decrease the false positive rate while increasing the false negative rate.

Turning to the cost, we have from above:

$$E[Cost] \approx \alpha P(A|\sim I)P(\sim I) + \beta P(\sim A|I)P(I)$$
$$= \alpha\left[1 - \Phi_{X|\mu_i=0}(x^\star)\right]P(\sim I) + \beta\Phi_{X|\mu_i=1}(x^\star)P(I)$$

And (in this example) can optimize with respect to cost by standard methods to find:

$$x^{\star}_{(opt)} = [\ln \alpha P(\sim I) - \ln \beta P(I)] + \frac{1}{2}$$

Or more generally, letting f_0 denote the density function for normal traffic and f_1 denote the density of anomalous traffic, we have the usual form of the weighted likelihood ratio decision rule[3]:

$$\frac{f_1(x^{\star})}{f_0(x^{\star})} = \frac{\beta P(I)}{\alpha P(\sim I)}$$

At this threshold $x^{\star}$, any further increase in $x^{\star}$ leads to an increase in the expected cost of analysis, as the benefit of reduced false positives in the left hand term begins to be outweighed by the cost of increased false negatives.

Critically, for many IDS deployments, $P(I)$ may be completely unknown, and β is generally only roughly approximated. This immediately suggests that increasing the threshold for an anomaly detector may in fact be counterproductive, and in many cases it is difficult or impossible to know when precisely this tradeoff has occurred.

While this is—as in [12]—generally grim news for anomaly detection, the impact of α and β on the cost is also of interest. Security policies, segregating sensitive resources, and controlling physical access to the most critical systems and information may provide avenues to reduce β, however we defer this analysis to others. The dependence on α, however, is worth examining. Notice that—again—it is a multiplicative factor to the rate of normal traffic, suggesting that small adjustments in α can have an impact significantly greater than an equal adjustment in β. Indeed, $\frac{d}{d\alpha}E[Cost] \approx P(A|\sim I)P(\sim I)$ while $\frac{d}{d\beta}E[Cost] = P(\sim A|I)P(I)$ where again $P(I) \ll P(\sim I)$. This suggests that reducing the cost to analysts of examining and diagnosing alarms may provide significant benefit with respect to the cost of operation of an IDS. As we have seen above, efforts to control the false positive rate in much of the academic work relating to IDSs have led to increasingly complex classification systems. We contend that this trend has in fact significantly increased α, greatly reducing the utility of such systems for practical use.

---

[3] Note another critical feature: if adversarial actors are capable of crafting their traffic to approximate f_0, such that the quantity $\left|1 - \frac{f_1(x)}{f_0(x)}\right| \leq \epsilon$ for some small $\epsilon > 0$, and can control the rate of malicious traffic they send and hence $P(I)$, then they may craft their traffic such that the defenders have no $x^{\star}$ that satisfies the above relationship and so cannot perform cost-effective anomaly detection. We do not discuss this problem in detail, but reserve it for future work.

5 Revisiting Tree-Based Anomaly Detectors

As discussed previously, tree-based anomaly IDSs were among the first considered [33]. The work of [33] focused largely on a host-based approach to detecting misuse of computer systems on the basis of host actions and made efforts to use the tree structure to extract rules that could be both automatically applied to audit trail logs as well as directly interpreted by analysts. While the majority of recent anomaly IDS work has focused on various very popular and highly successful kernel methods that have been developed for machine learning in other areas, other more recent approaches have leveraged the work that has been done in random decision forests [28, 29] and begun to investigate their application to both supervised anomaly detection [10, 30] and unsupervised outlier detection [31, 32] in both batched and online modes. The fact that tree-based ensemble classifiers can achieve extremely high performance without further transforming the data into a more complex space to make it easier to separate (although note the empirical observation in [41] that the best performance is often obtained by including additional transformations of the covariates) and are typically quite robust to ancillary data [29, 31] make them attractive targets for anomaly detection algorithms, and initial results [10, 30–32]—albeit often on limited data—have shown that their performance is comparable or even superior to methods that employ more complex transformations.

While the semantic gap is not widely addressed in these papers, trees naturally lend themselves to extracting contextual information and classification rules that are generally much more interpretable to end users than the distance-based metrics of kernel methods. They also handle the heterogeneous data observed in networks more naturally than many other classifiers which either explicitly operate on features that map naturally to inner product spaces (see virtually any KDD'99-based paper) or require some transformation of the data to convert it into such a space (e.g., [23], transforming sequential bytes into n-grams). In particular, the most common form of splitting rule within continuous covariates in trees—axis-aligned learners—allow us to extract simple inequalities to define portions of rules relating to those attributes, and splits of ordinal variables allow us to extract relevant subsets from such fields. By tracing a path from the root to the final leaf node of any tree, it is extremely straightforward to extract the "rule" that led to the classification of the point in question in that leaf node; tabulation of the number of data points in the training data that fell into each internal node or leaf immediately gives us conditional probabilities based that may be immediately extracted from the data. Figure 4 shows a toy example of a decision tree based on the West Point/ NSA CDX data [24] learned in an unsupervised manner similar to [32] (features selected at random from the available set, continuous features split at a randomly selected value from the ones available at that node, categorical features selected by information gain ratio; a maximum of three splits were permitted, and the splitting was terminated if the proposed split led to a leaf node of fewer than 100 points). A quick walk down the tree—taking the right-most split at each step—shows that we

Fig. 4 A toy decision tree

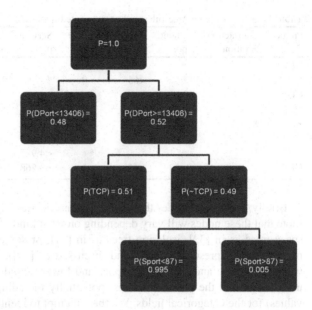

may construct the rule for an 'anomaly' learned by this tree as: "destination port > 13,406, not TCP, Source port > 87". This rule is presented using the type of data that analysts work with on a daily basis (rather than elaborate RKHS representations, for instance), and the value of the rule may be assessed immediately.[4]

While single trees are straightforward to assess, combining rules in multiple trees is less straightforward. This question was touched on in [33] in the context of pruning the extensive database of generated rules, but only for strictly categorical data. They employ several criteria for pruning, beginning with a threshold function on the quality of the rule (favoring longer, more accurate rules with a smaller range of acceptable consequents), then removing rules where the predicates form a simple permutation of some other rule, next removing rules where the consequent matches some other rule and the predicates form a special case of some other rule, and finally pruning on number of exemplars of the rule in the data and depth. These criteria can be easily transferred to the case of continuous covariates.

---

[4] In this case, any outgoing traffic to a relatively high destination port was deemed by an analyst to be unusual, but "certainly not a red flag"; the fact that it was non-TCP and did not originate from the lower end of the range of registered ports suggested a UDP streaming protocol, which often communicate across ephemeral ports; the analyst volunteered the suggestion that if it were in fact UDP it would likely not warrant further analysis. When the same analyst was presented with the outputs given in Fig. 1 through Fig. 3, they were of the opinion that it was not terribly useful, and that it not provide them with any guidance as to why it appeared suspicious; the semantic gap in action.

Table 1 Seven isolation tree rules and an associated meta-rule

Protocol	Transaction duration	Client port	Client packets	Client BPS	Service port	Service packets	Service BPS	$h(x)$
UDP	*	*	<69	*	*	*	*	2
*	*	*	<65	>0	*	*	*	2
UDP	*	*	*	*	*	*	>12,093	2
*	*	<23	*	*	*	*	>7,21,709	2
*	*	<137	*	*	*	*	>15,13,401	2
*	*	*	*	*	<3,268	>22	*	2
*	*	*	*	*	<4,985	*	>1,066	2
UDP	*	<23	<65	>0	<3,268	>22	>15,13,401	

Briefly, for all leaf nodes that lead to a classification of 'anomalous' in all trees (note that these nodes will vary depending on what kind of tree is used, either half-space trees as in [32], isolation trees as in [31], or some variety of density estimation tree as presented above and discussed in [29]), we extract the rules by walking the tree and identifying upper and lower bounds (if any) for the continuous fields, and the possible values (potentially including the special case of all values) for the categorical fields. We then attempt to identify the most compact set of aggregated rules for which the intersection of those sets is non-empty in all fields. As this process is equivalent to the set cover problem, we approximate the solution via a greedy algorithm, extracting the maximal set remaining in each iteration. Each aggregated rule is assigned a weight based on the number and weight of rules that are subsumed by it. In our example case, illustrated in Table 1, we used isolation trees of [31], simply taking their default values ($\psi = 256, t = 100, c(\psi) \approx 6.1123$). We can then define the weight of an aggregated rule as $s(x, \psi) = 2^{-\frac{\sum h_i(x)}{nc(\psi)}}$, as any packet that triggers our aggregated rule would, by definition, trigger all rules subsumed by the aggregated rule.

Note that the aggregated rule is extremely specific, represented in "natural units" of IDS data, and learned in an entirely unsupervised fashion from the data. In this case, analyst evaluation of the rule suggested that the primary feature of interest was the low port number in conjunction with the UDP protocol and the high rate of data from the external service. This suggested to the analyst a streaming protocol, consistent with many UDP transactions, but to a non-standard port. In addition, a feature that appeared to be irrelevant for this rule—the duration of the transaction—was naturally excluded, and could be easily trimmed from the display.

While this method takes an aggressive approach to rule pruning, essentially triggering only on instances that satisfy among the tightest constraints possible derived from the rule set represented by the trees, a similar approach could be used in a post hoc fashion, collecting only the subset of rules that the data point of interest triggered on its path through the forest, and performing a similar greedy aggregation method on them to determine aggregated rules that it satisfied. While this does not permit the one-time computation with subsequent amortization of the time complexity that pre-processing of the forest permits, the fixed number of trees in the ensemble mean that the system may operate, similarly to the method of [32], with constant time complexity (albeit potentially a large constant).

While broader-scale testing on realistic data is necessary, and additional work to identify the most useful features is required, preliminary results indicate that the approach of using tree-based anomaly detection systems to identify and extract human-readable rules describing the generated anomalies is of great promise, and suggests that anomaly detection systems may well be able cross the semantic gap and see some degree of deployability in settings beyond the research lab.

6 Conclusion

We present an overview of many current approaches in the literature to anomaly-based intrusion detection, and examine in detail the issue of the semantic gap first articulated by [1]. This semantic gap is due in large part to the emphasis of anomaly detection research on outlier detection methods and machine learning techniques that have been designed primarily to operate within inner product spaces, which are not in general representative of the problem space for actual IDS deployment and operation. The result of [12] shows that the reliability of an anomaly detector—the probability that investigating an alarm will lead to detection of malicious or anomalous activity—is directly related to the false positive rate of the detector, and so much emphasis has been placed on reducing this false positive rate when treating the IDS as a black box. We extend this result with a cost analysis to show that in operational deployment, the interpretability of an IDS alert is also significant, and suggest that tree-based detection methods that operate in the untransformed space of IDS data not only are better-suited to the IDS problem space, but also lend themselves more naturally to human interpretation than high-dimensional kernel-based methods without the sacrifices in accuracy that have characterized categorical outlier detection methods. We finally show a proof-of-concept method for extracting rules from ensembles of trees, producing novel rules for detecting anomalous traffic built in a completely unsupervised fashion from live network data.

Appendix A

Random decision tree classification of KDD'99 data was performed using the Scikit-learn [25] package under Python 2.7.2 on a desktop commodity workstation. Training was performed using the file kddcup.data_10_percent_corrected, and testing was done on the file kddcup.data.corrected. 494,021 training records were used, and 4,898,431 test records. The three fields "Count", "diff_srv_rate", and "dst_bytes" were extracted along with the label field in both data sets; all other

	A) Normal	Guess_passwd	Nmap	B) Loadmodule	Rootkit	Warezclient	C) Smurf	Pod	D) Neptune	Spy	ftp_write	Phf	E) Portsweep	Teardrop	Buffer_overflow	Land	Imap	F) Warezmaster	Perl	Multihop	Back	Ipsweep	G) Satan
A	0	53	2315	8	10	1020	971	264	5537	2	8	4	9558	752	28	20	12	5	3	7	2197	12480	950
B	1	0	0	0	0	0	0	0	0	0	0	0	0	0	1	0	0	0	0	0	0	0	0
C	2210	0	1	0	0	0	0	0	9672	0	0	0	1	119	1	1	0	0	0	0	0	0	95
D	402	0	0	0	0	0	40	0	0	0	0	0	42	108	0	0	0	0	0	0	0	1	970
E	0	0	0	0	0	0	0	0	0	0	0	0	0	0	0	0	0	0	0	0	0	0	0
F	3	0	0	0	0	0	0	0	0	0	0	0	3	0	0	0	0	0	0	0	0	0	3
G	6	0	0	0	0	0	0	0	0	0	0	0	0	0	0	0	0	0	0	0	0	0	0

Key: A) Normal B) Loadmodule C) Smurf D) Neptune E) Portsweep F) Warezmaster G) Satan data was discarded. The random decision forest was trained with the following parameters:

- Classification threshold: simple majority
- No bootstrapping used
- Features per node: 2
- Node splitting by informatin gain
- Minimum leaf samples: 1
- Minimum samples to split: 2
- Max tree depth: 9
- Number of trees: 11

Training the classifier required 4.4 s using a single processor, testing required approximately 122.8 s.[5] The following confusion matrix was produced (note that we have omitted correct classifications on the diagonal for compactness, and that we have also omitted rows corresponding to predictions that the classifier never produced).

Total false negatives: $36204/4898431 \approx 0.007$

Total false positives: $2622/4898431 \approx 0.0005$

The most common errors were misclassification of the IPsweep attack as normal traffic, and classification of flows corresponding to the Neptune attack as the Smurf attack. Random inspection of the IPsweep misclassifications suggests that each "attack" had several records associated with it; while many individual records were not correctly labeled, all instances that were examined by hand had at least one record in the total attack correctly classified. As the Smurf and Neptune attacks are both denial of service attacks, some confusion between the two is to be expected.

While these results certainly demonstrate that random decision forests are accurate and efficient classifiers, the alternative that the KDD'99 data is simply not a terribly representative data set for IDS research should not be excluded.

References

1. R. Sommer, V. Paxson, Outside the closed world: on using machine learning for network intrusion detection," in *2010 IEEE Symposium on Security and Privacy (SP)*, 2010
2. P. Laskov, P. DÃŒssel, C. SchÃ€fer, K. Rieck, in *Learning Intrusion Detection: Supervised or Unsupervised?*, ed. by F. Roli, S. Vitulano (Springer, Berlin, 2005), pp. 50–57
3. M. Roesch, Snort – lightweight intrusion detection for networks, in *Proceedings of the 13th USENIX Conference on System Administration*, 1999, pp. 229–238

---

[5] Due to the large size of the test set, it was not loaded into memory all at once, and instead was read sequentially from disk. Total time elapsed was 1023.3 s, of which profiling indicated that roughly 88 % was consumed by disk I/O operations. As our interest was in offhand comparison and not production use, we did not attempt to optimize this further.

4. V. Paxson, Bro: a system for detecting network intruders in real time. Comput. Netw. **31**(23–24), 2435–2463 (1999)
5. J. Long, D. Schwartz, S. Stoecklin, Distinguishing false from true alerts in Snort by data mining patterns of alerts, in *Proceedings of 2006 SPIE Defense and Security Symposium*, 2006
6. M. Sato, H. Yamaki, H. Takakura, Unknown attacks detection using feature extraction from anomaly-based IDS alerts, in *2012 IEEE/IPSJ 12th International Symposium on Applications and the Internet (SAINT)*, 2012
7. Y. Song, M.E. Locasto, A. Stavrou, A.D. Keromytis, S.J. Stolfo, On the infeasibility of modeling polymorphic shellcode – Re-thinking…, in *MACH LEARN*, 2009
8. H. Debar, M. Dacier, A. Wespi, Towards a taxonomy of intrusion-detection systems. Comput. Netw. **31**(8), 805–822 (1999)
9. O. Depren, M. Topallar, E. Anarim, M.K. Ciliz, An intelligent intrusion detection system (IDS) for anomaly and misuse detection in computer networks. Expert Syst. Appl. **29**(4), 713–722 (2005)
10. J. Zhang, M. Zulkernine, A. Haque, Random-forests-based network intrusion detection systems. IEEE Trans. Syst. Man Cybern. C Appl. Rev. **38**(5), 649–659 (2008)
11. N. Abe, B. Zadrozny, J. Langford, Outlier detection by active learning, in *Proceedings of the 12th ACM SIGKDD International Conference on Knowledge Discovery and Data Mining*, New York, 2006
12. S. Axelsson, The base-rate fallacy and the difficulty of intrusion detection. ACM Trans. Inf. Syst. Secur. **3**(3), 186–205 (2000)
13. A. Koufakou, E.G. Ortiz, M. Georgiopoulos, G.C. Anagnostopoulos, K.M. Reynolds, A scalable and efficient outlier detection strategy for categorical data, in *19th IEEE International Conference on Tools with Artificial Intelligence, 2007. ICTAI 2007*
14. M.E. Otey, A. Ghoting, S. Parthasarathy, Fast distributed outlier detection in mixed-attribute data sets. Data Min. Knowl. Discov. **12**(2–3), 203–228 (2006)
15. X. Song, M. Wu, C. Jermaine, S. Ranka, Conditional anomaly detection. IEEE Trans. Knowl. Data Eng. **19**(5), 631–645 (2007)
16. C. Cortes, V. Vapnik, Support-vector networks. Mach. Learn. **20**(3), 273–297 (1995)
17. K. Wang, S. Stolfo, One-class training for Masquerade detection, in *Workshop on Data Mining for Computer Security*, 2003
18. R. Perdisci, G. Gu, W. Lee, Using an ensemble of one-class SVM classifiers to harden payload-based anomaly detection systems, in *Sixth International Conference on Data Mining, 2006. ICDM'06*. 2006
19. S. Mukkamala, G. Janoski, A. Sung, Intrusion detection using neural networks and support vector machines, in *Proceedings of the 2002 International Joint Conference on Neural Networks*, 2002
20. J. Weston, C. Watkins, Technical Report CSD-TR-98-04, Department of Computer Science, *Multi-class Support Vector Machines*, Royal Holloway, University of London, 1998
21. R. Chen, K. Cheng, Y. Chen, C. Hsieh, Using rough set and support vector machine for network intrusion detection system, in *First Asian Conference on Intelligent Information and Database Systems*, 2009
22. T. Shon, Y. Kim, C. Lee, J. Moon, A machine learning framework for network anomaly detection using SVM and GA, in *Information Assurance Workshop, 2005. IAW'05. Proceedings from the Sixth Annual IEEE SMC*, 2005
23. K. Wang, S. Stolfo, Anomalous payload-based network intrusion detection, in *Recent Advances in Intrusion Detection*, 2004
24. B. Sangster, T. O'Connor, T. Cook, R. Fanelli, E. Dean, J. Adams, C. Morrell, G. Conti, Toward instrumenting network warfare competitions to generate labeled datasets, in *USENIX Security's Workshop on Cyber Security Experimentation and Test (CSET)*, 2009
25. F. Pedregosa, G. Varoquaux, A. Gramfort, V. Michel, B. Thirion, O. Grisel, M. Blondel, P. Prettenhofer, R. Weiss, V. Dubourg, J. Vanderplas, A. Passos, D. Cournapeau, M. Brucher,

M. Perrot, E. Duchesnay, Scikit-learn: machine learning in python. J. Mach. Learn. Res. **12**, 2825–2830 (2011)

26. P. Biondi, Scapy, a powerful interactive packet manipulation program. , *Scapy*, 2011, http://www.secdev.org/projects/scapy/
27. V. Frias-Martinez, J. Sherrick, S.J. Stolfo, A.D. Keromytis, A network access control mechanism based on behavior profiles, in *Computer Security Applications Conference, 2009. ACSAC'09. Annual*, 2009
28. L. Breiman, Random forests. Mach. Learn. **45**(1), 5–32 (2001)
29. A. Criminisi, J. Shotton, E. Konukoglu, *Decision Forests for Classification, Regression, Density Estimation, Manifold Learning and Semi-Supervised Learning*, Microsoft Technical Report, 2011
30. D.S. Kim, S.M. Lee, J.S. Park, *Building Lightweight Intrusion Detection System Based on Random Forest*, ed. by J. Wang, Z. Yi, J.M. Zurada, B. Lu, H. Yin (Springer, Berlin, 2006), pp. 224–230
31. F.T. Liu, K.M. Ting, Z.-H. Zhou, Isolation-based anomaly detection. ACM Trans. Knowl. Discov. Data **6**(1), 3:1–3:39 (2012)
32. S.C. Tan, K.M. Ting, T.F. Liu, Fast anomaly detection for streaming data, in *Proceedings of the Twenty-Second International Joint Conference on Artificial Intelligence*, vol. 2, 2011
33. H.S. Vaccaro, G.E. Liepins, Detection of anomalous computer session activity, in *Proceedings of 1989 IEEE Symposium on Security and Privacy*, 1989
34. D.E. Denning, An intrusion-detection model. IEEE Trans. Softw. Eng. **13**(2), 222–232 (1987)
35. M. Mahoney, P. Chan, An analysis of the 1999 DARPA/Lincoln laboratory evaluation data for network anomaly detection, in *Recent Advances in Intrusion Detection*, 2003
36. J. McHugh, Testing intrusion detection systems: a critique of the 1998 and 1999 DARPA intrusion detection system evaluations as performed by Lincoln Laboratory. ACM Trans. Inf. Syst. Secur. **3**(4), 262–294 (2000)
37. T. Lunt, A. Tamaru, F. Gilham, R. Jagannathan, C. Jalali, P. Neumann, H. Javitz, A. Valdes, T. Garvey, *A real-time intrusion-detection expert system (IDES)*, SRI International, Computer Science Laboratory, 1992
38. M. Molina, I. Paredes-Oliva, W. Routly, P. Barlet-Ros, Operational experiences with anomaly detection in backbone networks. Comput. Secur. **31**(3), 273–285 (2012)
39. K.M. Tan, R.A. Maxion, "Why 6?" Defining the operational limits of stide, an anomaly-based intrusion detector, in *Proceedings of the IEEE Symposium on Security and Privacy*, 2001
40. L. Sassaman, M.L. Patterson, S. Bratus, A. Shubina, The Halting problems of network stack insecurity, in *USENIX*, 2011
41. Z. Zhou, *Ensemble Methods: Foundations and Algorithms* (Chapman & Hall, 2012)

Recognizing Unexplained Behavior in Network Traffic

Massimiliano Albanese, Robert F. Erbacher, Sushil Jajodia, C. Molinaro, Fabio Persia, Antonio Picariello, Giancarlo Sperlì and V. S. Subrahmanian

1 Introduction

Intrusion detection and alert correlation techniques provide valuable and complementary tools for identifying and monitoring security threats in complex network infrastructures. Intrusion detection systems (IDS) can monitor network traffic for suspicious behavior and trigger security alerts accordingly [1–3]. Alert correlation methods can aggregate such alerts into multi-step attack scenarios [4–8].

Intrusion detection has been studied for about 30 years, since it was first identified in the Anderson report [9], and it is based on the assumption that an

M. Albanese (✉) · S. Jajodia
George Mason University, Fairfax, VA, USA
e-mail: malbanes@gmu.edu

S. Jajodia
e-mail: jajodia@gmu.edu

R. F. Erbacher
US Army Research Laboratory, Adelphi, MD, USA
e-mail: robert.f.erbacher.civ@mail.mil

C. Molinaro
University of Calabria, Rende, Italy
e-mail: cmolinaro@deis.unical.it

F. Persia · A. Picariello · G. Sperlì
University of Naples Federico II, Naples, Italy
e-mail: fabio.persia@unina.it

A. Picariello
e-mail: picus@unina.it

G. Sperlì
e-mail: g.sperli@unina.it

V. S. Subrahmanian
University of Maryland, College Park, MD, USA
e-mail: vs@umiacs.umd.edu

R. E. Pino (ed.), *Network Science and Cybersecurity*,
Advances in Information Security 55, DOI: 10.1007/978-1-4614-7597-2_3,
© Springer Science+Business Media New York 2014

intruder's behavior will be noticeably different from that of a legitimate user and that many unauthorized actions are detectable [3].

Intrusion detection techniques can be broadly classified into *signature-based* [2] and *profile-based* (or *anomaly-based*) [1] methods. There are advantages and disadvantages to each method. The trend today is to use the two methods together to provide the maximum defense for the network infrastructure. A signature generally refers to a set of conditions that characterize the direct manifestation of intrusion activities in terms of packet headers and payload content. Historically, signature-based methods have been most common when looking for suspicious or malicious activity on the network. These methods rely on a database of attack signatures and when one or more of these signatures match what is observed in live traffic, an alarm is triggered and the event is logged for further investigation. The effectiveness of signature-based intrusion detection is highly dependent on its signature database. If a signature is not in the database, the IDS will not recognize the attack.

Anomaly-based intrusion detection triggers an alarm when some type of unusual behavior occurs on the network. This would include any event that is considered to be abnormal by a predefined standard. Anything that deviates from this baseline of *normal* behavior will be flagged and logged as *anomalous* or *suspicious*. For instance, HTTP traffic on a non-standard port, say port 63, would be flagged as suspicious. Normal behavior can be programmed into the system based on offline learning and research, or the system can learn the normal behavior online while processing the network traffic.

In complex networks, most intrusions are not isolated but represent different stages of specific attack sequences, with the early stages preparing for the later ones. In other words, complex attacks consist of multiple steps, each triggering specific security alerts. This fundamental observation, along with the potentially large number of alerts deriving from the widespread deployment of IDS sensors, has prompted significant research in automatic alert correlation techniques. The goal of correlation is to find causal relationships between alerts in order to reconstruct attack scenarios from isolated alerts [10]. Although it may not significantly reduce the number of alerts, the main role of correlation is to provide a higher level view of the actual attacks. Existing approaches to alert correlation can be divided into the following categories based on the criterion used for relating alerts: scenario-based correlation [4, 5], rule-based correlation [6], statistical correlation [11, 12], and temporal correlation [13].

From a conceptual point of view, both intrusion detection systems and alert correlation methods aggregate fine grained information into higher level views of the attack, although they operate at different levels of abstraction, as shown in Fig. 1. Moreover, both rely on models encoding a priori knowledge of either normal or malicious behavior, and cannot appropriately deal with events that are not *explained* by the underlying models. In practice, all these methods are incapable of quantifying how well available models explain a sequence of events observed in data streams (data packets and alerts respectively) feeding the two classes of tools.

Fig. 1 Conceptual diagram of the relationship between alert correlation and intrusion detection

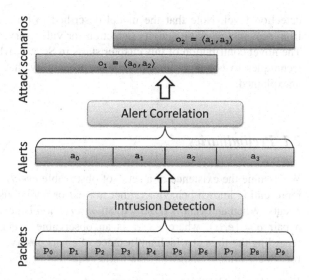

To address this limitation and offer novel analytic capabilities, we present a framework for evaluating the probability that a sequence of events—either at the network traffic level or at the alert level—is unexplained, given a set of models of previous learned behaviors. Our approach is an application of the framework proposed in [14] to the cyber security setting. We adapt algorithms from [14] so as to efficiently *estimate* the probability that a sequence is unexplained (rather than computing the exact probability as done in [14]). The computation of approximate probabilities is done by leveraging the mathematical properties studied in Sect. 3. Our framework can operate both at the intrusion detection level and at the alert correlation level, but it is not intended to replace existing tools. In fact, our framework builds on top of these tools, and analyzes their output in order to identify what is not "sufficiently" explained by the underlying models. Experiments on a prototype implementation of the framework show that our approach scales well and provides accurate results.

The rest of the chapter is organized as follows. Section 2 presents the proposed probabilistic model, whereas Sect. 3, discusses the properties that can be leveraged to compute approximate probabilities efficiently. Efficient algorithms to recognize unexplained behaviors are presented in Sect. 4. An experimental evaluation of our framework is reported in Sect. 5, and concluding remarks are given in Sect. 6.

2 Behavior Model

In this section, we present a framework for evaluating the probability that a sequence of events is unexplained, given a set of models. As already mentioned above, this framework can operate both at alert correlation level and at intrusion detection level. Note that the model described in this section has been adapted from previous work on activity detection for video surveillance applications [14]. The novel contribution of this chapter starts in Sect. 3, where we propose efficient techniques to compute an approximate probability that a sequence of events is unexplained.

2.1 Preliminaries

We assume the existence of a set $\mathscr{E}$ of observable *events*,[1] and a set $\mathscr{A}$ of *models*[2] representing known behavior (either normal or malicious) in terms of observable events. When an event is observed, an *observation* is generated. An observation is a pair $a = (e, ts)$, where $e \in \mathscr{E}$ is an observable event, and ts is a timestamp recording the time at which an instance of e was observed. An *observation stream* (or *observation sequence*) S is a finite sequence of observations.

Example 1 Consider the Snort rule `alert any any → any any` `(flags:SF,12; msg:``Possible SYN FIN scan'';)`.[3] This rule detects when a packet has the SYN and FIN flags set at the same time—indicating a possible SYN FIN scan attempt—and generates an alert (observation) $a = (e, ts)$, where the observable event e is the fact that the SYN and FIN flags are set at the same time, and ts is the time at which the packet was observed.

Throughout the chapter, we use the following terminology and notation for sequences. Let $S_1 = \langle a_1, \ldots, a_n \rangle$ and $S_2 = \langle b_1, \ldots, b_m \rangle$ be two sequences. We say that S_2 is a *subsequence* of S_1 iff there exist $1 \leq j_1 < j_2 < \cdots < j_m \leq n$ s.t. $b_i = a_{j_i}$ for $1 \leq i \leq m$. If $j_{i+1} = j_i + 1$ for $1 \leq i < m$, then S_2 is a *contiguous* subsequence of S_1. We write $S_1 \cap S_2 \neq \emptyset$ iff S_1 and S_2 have a common element and write $e \in S_1$ iff e is an element appearing in S_1. The *concatenation* of S_1 and S_2, i.e., the sequence $\langle a_1, \ldots, a_n, b_1, \ldots, b_m \rangle$, is denoted by $S_1 \cdot S_2$. Finally, we use $|S_1|$ to denote the length of S_1, that is, the number of elements in S_1.

Given an observation stream S, and a behavior model $A \in \mathscr{A}$, an *occurrence* o of A in S is a subsequence $\langle (e_1, ts_1), \ldots (e_m, ts_m) \rangle$ of S such that the sequence of events $e_1, \ldots, e_m$ represents a possible way of exhibiting behavior A (e.g., a specific path in the attack graph from initial to target conditions). The relative probability[4] $p^*(o)$ of the occurrence o is the probability that the sequence of events

---

[1] At the intrusion detection level, observable events may simply be observable packet features. At the alert correlation level, observable events are alerts generated by the underlying intrusion detection system.

[2] At the intrusion detection level, $\mathscr{A}$ is a set of IDS rules. At the alert correlation level, $\mathscr{A}$ is a set of attack models, such as attack graphs.

[3] http://www.snort.org/.

[4] Probabilities of occurrences must be normalized in order to enable comparison of occurrences of different behavior models.

$\langle e_1, \ldots, e_m \rangle$ is in fact an instance of the corresponding behavior. The problem of computing the probability of an occurrence is beyond the scope of our work. However, several researchers have addressed this problem. For instance, a probabilistic extension of attack graphs is proposed in [15], along with an algorithm for identifying attack occurrences efficiently. Therefore, we assume that all the occurrences o of a behavior A and their probabilities can be readily computed.

We use $\mathcal{O}(S, \mathcal{A})$ to denote the set of all occurrences in S of behaviors in $\mathcal{A}$. When S and $\mathcal{A}$ are clear from the context, we write $\mathcal{O}$ instead of $\mathcal{O}(S, \mathcal{A})$.

2.2 Probabilistic Unexplained Behavior Model

We now define the probability that an observation sequence is unexplained, given a set $\mathcal{A}$ of known behaviors. We start by noting that the probabilistic nature of *occurrences* implicitly involves conflicts. For instance, consider the two occurrences o_1 and o_2 in Fig. 2. In this case, there is an implicit conflict because a_2 belongs to both occurrences, but in fact, a_2 can only belong to one occurrence, i.e., though o_1 and o_2 may both have a non-zero probability of occurrence, the probability that they coexist is $0.^5$ Formally, we say two occurrences o_i, o_j *conflict*, denoted $o_i \approx o_j$, iff $o_i \cap o_j \neq \emptyset$. We now use this notion to define possible worlds.

Definition 1 (Possible world) A *possible world* for an observation sequence S w.r.t. a set of behavior models $\mathcal{A}$ is a subset w of $\mathcal{O}(S, \mathcal{A})$ s.t. $\nexists o_i, o_j \in w, o_i \approx o_j$.

Thus, a possible world is a set of occurrences which do not conflict with one another, i.e., an observation cannot be identified as part of two distinct occurrences in the same world. We use $\mathcal{W}(S, \mathcal{A})$ to denote the set of all possible worlds for an observation sequence S w.r.t. a set of behavior models $\mathcal{A}$; when S and $\mathcal{A}$ are clear from context, we write $\mathcal{W}$.

Example 2 Consider the observation sequence and the two conflicting occurrences o_1, o_2 in Fig. 2. There are three possible worlds: $w_0 = \emptyset$, $w_1 = \{o_1\}$, and $w_2 = \{o_2\}$. Note that $\{o_1, o_2\}$ is not a world as $o_1 \approx o_2$. Each world represents a way of explaining what is observed. The first world corresponds to the case where nothing is explained, the second and third worlds correspond to the scenarios where we use one of the two possible occurrences to explain the observed events.

Note that any subset of $\mathcal{O}$ not containing conflicting occurrences is a legitimate possible world—possible worlds are not required to be maximal w.r.t. set inclusion $\subseteq$. In the above example, the empty set is a possible world even though there are two other possible worlds $w_1 = \{o_1\}$ and $w_2 = \{o_2\}$ which are supersets of it.

---

[5] This assumption makes modeling simpler, but it can be removed or modified in situations where certain atomic events are shared among multiple attack patterns.

Fig. 2 Example of
conflicting occurrences

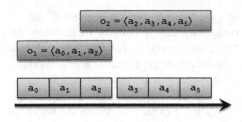

The reason is that o_1 and o_2 are uncertain, so the scenario where neither o_1 nor o_2 occurs is a legitimate one. We further illustrate this point below.

Example 3 Consider an observation sequence where a single occurrence o has been identified with $p^*(o) = 0.6$. In this case, it is natural to say that there are two possible worlds $w_0 = \emptyset$ and $w_1 = \{o\}$ and expect the probabilities of w_0 and w_1 to be 0.4 and 0.6, respectively. By restricting ourselves to maximal possible worlds only, we would have only one possible world, w_1, whose probability is 1, which is wrong. Nevertheless, if $p^*(o) = 1$, w_1 is the only possible scenario. This can be achieved by assigning 0 and 1 to the probabilities of w_0 and w_1, respectively.

Thus, occurrences determine a set of possible worlds (intuitively, different ways of explaining the observation stream). We wish to find a probability distribution over all possible worlds that (1) is consistent with the relative probabilities of the occurrences, and (2) takes conflicts into account. We assume the user specifies a function $\omega : \mathscr{A} \to \mathbb{R}^+$ which assigns a weight to each behavior and prioritizes the importance of the behavior.[6] The weight of an occurrence o of behavior A is the weight of A.

We use $C(o)$ to denote the set of occurrences conflicting with o, i.e., $C(o) = \{o' | o' \in \mathcal{O} \wedge o' \nsim o\}$. Note that $o \in C(o)$; furthermore, $C(o) = \{o\}$ when o does not conflict with any other occurrence. Suppose p_i denotes the (unknown) probability of world w_i. As we know the probability of occurrences, and as each occurrence occurs in certain worlds, we can induce a set of linear constraints that can be used to learn the values of the p_i's.

Definition 2 Let S be an observation sequence, $\mathscr{A}$ a set of behavior models, and $\mathcal{O}$ the set of occurrences identified in S w.r.t. $\mathscr{A}$. We define the linear constraints $LC(S, \mathscr{A})$ as follows:

$$
\begin{cases}
p_i \geq 0, \forall w_i \in \mathscr{W} \\
\sum_{w_i \in \mathscr{W}} p_i = 1 \\
\sum_{w_i \in \mathscr{W} \text{ s.t.} o \in w_i} p_i = p^*(o) \cdot \frac{\omega(o)}{\sum_{o_j \in C(o)} \omega(o_j)}, \forall o \in \mathcal{O}
\end{cases}
$$

---

[6] For instance, highly threatening behaviors may be assigned a high weight.

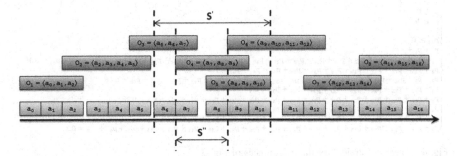

Fig. 3 Example of observation sequence and occurrences

Table 1 Probabilities and weights of occurrences	o	$p^*(o)$	$\omega(o)$	$p^*(o) \cdot \dfrac{\omega(o)}{\sum_{o_j \in C(o)} \omega(o_j)}$
	o_1	0.90	3	0.45
	o_2	0.80	3	0.30
	o_3	0.72	2	0.16
	o_4	0.65	4	0.20
	o_5	0.77	4	0.28
	o_6	0.50	3	0.10
	o_7	0.60	4	0.24
	o_8	0.70	3	0.30

The first two types of constraints enforce a probability distribution over the set of possible worlds. The last type of constraint ensures that the probability of occurrence o—which is the sum of the probabilities of the worlds containing o—is equal to its relative probability $p^*(o)$ weighted by $\dfrac{\omega(o)}{\sum_{o_j \in C(o)} \omega(o_j)}$, the latter being the weight of o divided by the sum of the weights of the occurrences conflicting with o. Note that: (i) the value on the right-hand side of the last type of constraint decreases as the amount of conflict increases, (ii) if an occurrence o is not conflicting with any other occurrence, then its probability $\sum_{w_i \in \mathcal{W} \text{ s.t.} o \in w_i} p_i$ is equal to $p^*(o)$, i.e., the relative probability returned by chosen tool for identifying behavior occurrences in observation streams.

Example 4 Consider the observation sequence and occurrences of Fig. 3, and assume that such occurrences have been identified with the relative probabilities shown in the second column of Table 1. The table also shows the weights assigned to the occurrences and the value of $p^*(o) \cdot \dfrac{\omega(o)}{\sum_{o_j \in C(o)} \omega(o_j)}$. The 8 occurrences determine 49 possible worlds.[7] The set of linear constraints of Definition 2 for this case is shown in Fig. 4.

---

[7] We do not list all the worlds for reason of space.

$$
\begin{cases}
p_0 \geq 0 \\
p_1 \geq 0 \\
\ldots \\
p_{48} \geq 0 \\
p_0 + p_1 + \cdots + p_{48} = 1 \\
p_1 + p_9 + p_{10} + p_{11} + p_{12} + p_{13} + p_{14} + p_{28} + p_{29} + p_{30} + p_{31} + p_{32} + p_{33} + p_{34} + p_{35} + p_{36} + p_{45} + p_{46} + p_{47} + p_{48} = 0.45 \\
p_2 + p_{15} + p_{16} + p_{17} + p_{18} + p_{19} + p_{38} + p_{39} + p_{40} + p_{41} + p_{42} = 0.3 \\
p_3 + p_9 + p_{20} + p_{21} + p_{22} + p_{23} + p_{29} + p_{30} + p_{31} + p_{32} + p_{43} + p_{44} + p_{45} + p_{46} + p_{47} + p_{48} = 0.16 \\
p_4 + p_{10} + p_{15} + p_{24} + p_{25} + p_{33} + p_{34} + p_{38} + p_{39} = 0.2 \\
p_5 + p_{11} + p_{16} + p_{20} + p_{26} + p_{27} + p_{29} + p_{35} + p_{36} + p_{40} + p_{41} + p_{43} + p_{44} + p_{46} + p_{47} = 0.28 \\
p_6 + p_{12} + p_{17} + p_{21} + p_{28} + p_{30} + p_{37} + p_{42} + p_{45} + p_{48} = 0.1 \\
p_7 + p_{13} + p_{18} + p_{22} + p_{24} + p_{26} + p_{31} + p_{33} + p_{35} + p_{38} + p_{40} + p_{43} + p_{46} = 0.24 \\
p_8 + p_{14} + p_{19} + p_{23} + p_{25} + p_{27} + p_{28} + p_{32} + p_{34} + p_{36} + p_{37} + p_{39} + p_{41} + p_{42} + p_{44} + p_{45} + p_{47} + p_{48} = 0.3
\end{cases}
$$

Fig. 4 Linear constraints for the occurrences of Fig. 3

In the rest of the chapter, we assume that $LC(S, \mathscr{A})$ is solvable. We give two semantics for a subsequence S' of an observation sequence S to be unexplained in a world $w \in \mathscr{W}$:

1. S' is *totally unexplained* in w, denoted $w \not\models_T S'$, iff $\forall a_i \in S', \nexists o \in w, a_i \in o$;

2. S' is *partially unexplained* in w, denoted $w \not\models_P S'$, iff $\exists a_i \in S', \nexists o \in w, a_i \in o$.

Intuitively, S' is totally (resp. partially) unexplained in w iff w does not explain every (resp. at least one) observation in S'.

When we have a probability distribution over the set of possible worlds (i.e., a solution of $LC(S, \mathscr{A})$), the probability that a sequence S' is totally (resp. partially) unexplained can be naturally defined as the sum of the probabilities of the worlds w_i s.t. S' is totally (resp. partially) unexplained in w_i. This is formally defined as follows.

Definition 3 Let $\mathscr{A}$ be a set of behavior models, S an observation sequence, and S' a subsequence of S. Suppose we have a probability distribution ϕ over $\mathscr{W}$ obtained by solving $LC(S, \mathscr{A})$. The probability that S' is totally unexplained in S w.r.t. $\mathscr{A}$ and ϕ is

$$
\mathscr{P}_T(S', \mathscr{A}, \phi) = \sum_{w_i \in \mathscr{W} \text{ s.t. } w_i \not\models_T S'} \phi(w_i)
$$

Similarly, the probability that S' is partially unexplained in S w.r.t. $\mathscr{A}$ and ϕ is

$$
\mathscr{P}_P(S', \mathscr{A}, \phi) = \sum_{w_i \in \mathscr{W} \text{ s.t. } w_i \not\models_P S'} \phi(w_i)
$$

The previous definition gives the probability that a sequence S' is totally (resp. partially) unexplained for a given solution of $LC(S, \mathscr{A})$. However, in general $LC(S, \mathscr{A})$ can admit multiple solutions, each yielding a probability that a sequence is totally or partially unexplained. We define the probability interval that a sequence S' is totally (resp. partially) unexplained by minimizing and maximizing the probability that S' is totally (resp. partially) unexplained subject to the linear constraints of Definition 2.

Definition 4 Let $\mathscr{A}$ be a set of behavior models, S an observation sequence, and S' a subsequence of S. The probability interval that S' is totally unexplained in S w.r.t. $\mathscr{A}$ is $\mathscr{I}_T(S', \mathscr{A}) = [l, u]$, where:

$$l = \textbf{minimize} \sum\nolimits_{w_i \in \mathscr{W} \text{ s.t. } w_i \not\models_T S'} p_i$$

$$\textbf{subject to } LC(S, \mathscr{A})$$

$$u = \textbf{maximize} \sum\nolimits_{w_i \in \mathscr{W} \text{ s.t. } w_i \not\models_T S'} p_i$$

$$\textbf{subject to } LC(S, \mathscr{A})$$

Likewise, the probability interval that S' is partially unexplained in S w.r.t. $\mathscr{A}$ is $\mathscr{I}_P(S', \mathscr{A}) = [l', u']$, where:

$$l' = \textbf{minimize} \sum\nolimits_{w_i \in \mathscr{W} \text{ s.t. } w_i \not\models_P S'} p_i$$

$$\textbf{subject to } LC(S, \mathscr{A})$$

$$u' = \textbf{minimize} \sum\nolimits_{w_i \in \mathscr{W} \text{ s.t. } w_i \not\models_P S'} p_i$$

$$\textbf{subject to } LC(S, \mathscr{A})$$

Thus, the probability $\mathscr{P}_T(S', \mathscr{A}, \phi)$ (resp. $\mathscr{P}_P(S', \mathscr{A}, \phi)$) that a subsequence S' of S is totally (*resp.* partially) unexplained w.r.t. a solution ϕ of $LC(S, \mathscr{A})$ is the sum of the probabilities of the worlds in which S' is totally (resp. partially) unexplained. As $LC(S, \mathscr{A})$ may have multiple solutions, we find the tightest interval $[l, u]$ (resp. $[l', u']$) containing this probability for any solution. Different criteria can be used to choose a point probability value from an interval $[l, u]$, e.g., the minimum (l), the maximum (u), or the average (i.e., $(l + u)/2$).

In the rest of the chapter, we assume that one of the above criteria has been chosen, and we use $\mathscr{P}_T(S', \mathscr{A})$ (resp. $\mathscr{P}_P(S', \mathscr{A})$) to denote the probability that S' is totally (resp. partially) unexplained; when $\mathscr{A}$ is clear from context, we write $\mathscr{P}_T(S')$ (resp. $\mathscr{P}_P(S')$).

Example 5 Consider the observation sequence and occurrences of Fig. 3. The probability $\mathscr{P}_T(S')$ that the sequence $S' = \langle a_6, a_7, a_8, a_9, a_{10} \rangle$ is totally unexplained is obtained by minimizing and maximizing the objective function $\sum_{w_i \in Ws.t. w \not\models_T S'} p_i = p_0 + p_1 + p_2 + p_7 + p_8 + p_{13} + p_{14} + p_{18} + p_{19}$ subject to the constraints of Fig. 4, which gives $\mathscr{I}_T(S') = [0.26, 0.42]$, that is $0.26 \leq \mathscr{P}_T(S') \leq 0.42$. The probability $\mathscr{P}_P(S')$ that the sequence $S'' = a_7, a_8$ is partially unexplained is obtained by minimizing and maximizing the corresponding objective function[8] which gives $\mathscr{I}_P(S'') = [0.64, 0.8]$.

Proposition 1 *Consider two subsequences S_1 and S_2 of an observation sequence S. If S_1 is a subsequence of S_2, then $\mathscr{P}_T(S_1) \geq \mathscr{P}_T(S_2)$ and $\mathscr{P}_P(S_1) \leq \mathscr{P}_P(S_2)$.*

We now define totally and partially unexplained behaviors.

Definition 5 (Unexplained behavior) Let S be an alert sequence, $\tau \in [0, 1]$ a probability threshold, and $L \in \mathbb{N}^+$ a length threshold. Then,

- *a totally unexplained behavior* is a subsequence S' of S s.t. (i) $\mathscr{P}_T(S') \geq \tau$, (ii) $|S'| \geq L$, and (iii) S' is maximal, i.e., there does not exist a subsequence $S'' \neq S'$ of S s.t. S' is a subsequence of S'', $\mathscr{P}_T(S'') \geq \tau$, and $|S''| \geq L$.
- *a partially unexplained behavior* is a subsequence S' of S s.t. (i) $\mathscr{P}_P(S') \geq \tau$, (ii) $|S'| \geq L$, and (iii) S' is minimal, i.e., there does not exist a subsequence $S'' \neq S'$ of S s.t. S'' is a subsequence of S', $\mathscr{P}_P(S'') \geq \tau$, and $|S''| \geq L$.

In the definition above, L is the minimum length a sequence must be for it to be considered a possible unexplained behavior. Totally unexplained behaviors (TUBs for short) S' have to be maximal because, based on Proposition 1, any subsequence of S' is totally unexplained with probability greater than or equal to that of S'. On the other hand, partially unexplained behaviors (PUBs for short) S' have to be minimal because, based on Proposition 1, any super-sequence of S' is partially unexplained with probability greater than or equal to that of S'.

Intuitively, an unexplained behavior is a sequence of events that are observed on a network and poorly explained by known behavior models. Such sequences might correspond to unknown variants of known behaviors or to entirely new—and unknown—behaviors. As such, the proposed approach may be help in discovering zero-day attacks, which are unknown to administrators by definition.

An *Unexplained Behavior Problem* (UBP) instance is a four-tuple $I = \langle S, \mathscr{A}, \tau, L \rangle$, where S is an alert sequence, $\mathscr{A}$ is a set of behavior models, $\tau \in [0, 1]$ is a probability threshold, and $L \in \mathbb{N}^+$ is a length threshold. We want to find the sets $\mathcal{O}^{tu}(I)$ and $\mathcal{O}^{pu}(I)$ of all totally and partially unexplained behaviors in S, respectively.

The unexplained behavior model presented in this section applies to both alert correlation and intrusion detection. When used at the intrusion detection level, observable events are packet features and models are IDS rules. When used at the alert correlation level, observable events are IDS alerts and models are attack models, such as attack graphs.

---

[8] This objective function is the sum of 34 variables and is not shown for reasons of space.

3 Properties

Previous work on the recognition of unexplained activities [14] relies on an independence assumption to break the large optimization problem of Definition 4 into smaller optimization problems.[9] Specifically, [14] uses the transitive closure of $\sim$ to determine a partition of the set $\mathcal{O}$ of activity occurrences into equivalence classes $\mathcal{O}_1, \ldots, \mathcal{O}_m$, and assume that activity occurrences in one class are independent of activity occurrences in another class. Although this assumption is reasonable in the realm of video data, where periods of low or no activity in the field of view of a single camera are likely to break the flow of events into independent segments, we drop such an assumption for the purpose of identifying unexplained behaviors in network intrusions. In fact, an observation stream typically includes alerts from multiple sources, and multiple activities may be occurring at any given time, making conflict based partitioning ineffective. For example, conflict based partitioning of the set of occurrences in Fig. 3 leads to a single equivalence class containing all the occurrences.

In this section, we derive properties that can be leveraged to solve UBPs efficiently.

3.1 Totally Unexplained Behaviors

First, given a sequence S', we show that lower and upper bounds for $\mathscr{P}_T(S')$ can be found without solving the optimization problem of Definition 4. In order to do so, we introduce the following preliminary definition.

Definition 6 (Maximal intersecting set of occurrences) Let $\mathcal{O}^*$ be a set of occurrences. A *maximal intersecting set of occurrences* for $\mathcal{O}^*$ is a subset $\mathcal{O}'$ of $\mathcal{O}^*$ such that:

- $\forall o_i, o_j \in O', o_i \sim o_j$; and
- $\nexists \mathcal{O}' \subseteq \mathcal{O}^* \text{s.t.} \mathcal{O}' \subset \mathcal{O}'' \land \forall o_i, o_j \in O'', o_i \sim o_j$;

Intuitively, a set of occurrences is *intersecting* iff any two occurrences in the set conflict. An intersecting set of occurrences is *maximal* iff no proper superset of it is an intersecting set.[10] We use $\mathscr{M}(\mathcal{O}^*)$ to denote the set of maximal intersecting sets of occurrences in $\mathcal{O}^*$.

---

[9] Indeed, the set of constraints becomes non-linear with the addition of the constraints reflecting the independence assumption.

[10] The problem of finding all the maximal intersecting sets of occurrences is a generalization of the problem of finding maximal intersecting families of k-sets, but it is more general as occurrences are not required to have the same length k. As we need to compute maximal intersecting sets for small sets $\mathcal{O}^*$ of occurrences, complexity of this problem is not an issue.

Example 6 Consider the observation sequence of Fig. 3, and let $\mathcal{O}$ be the set of all occurrences recognized in the sequence. The set $\{o_4, o_5, o_6\}$ is a maximal intersecting set of occurrences for $\mathcal{O}$, as $o_4 \sim o_5$, $o_4 \sim o_6$, and $o_5 \sim o_6$, and there is no proper superset containing pairwise conflicting occurrences. Instead, the set $\{o_3, o_4, o_5\}$ is not a maximal intersecting set of occurrences because o_3 and o_5 do not conflict. In this case, the set of all maximal intersecting sets of occurrences in $\mathcal{O}$ is $\mathcal{M}(\mathcal{O}) = \{\{o_1, o_2\}, \{o_2, o_3\}, \{o_3, o_4\}, \{o_4, o_5, o_6\}, \{o_6, o_7\}, \{o_7, o_8\}\}$.

Theorem 1 *Consider a subsequence S' of an observation sequence S and the set $\mathcal{O}$ of occurrences identified in S w.r.t. a set $\mathcal{A}$ of behavior models, and let $\mathcal{O}^*$ be the set of occurrences $o \in \mathcal{O}$ such that $o \cap S' \neq \emptyset$. Then*

$$\mathscr{P}_T\left(S'\right) \geq 1 - \min\left\{1, \sum_{o \in \mathcal{O}^*} p^*(o) \cdot \frac{\omega(o)}{\sum_{o_j \in C(o)} \omega(o_j)}\right\} \tag{1}$$

$$\mathscr{P}_T\left(S'\right) \leq 1 - \max_{\mathcal{O}' \in \mathcal{M}(\mathcal{O}^*)} \sum_{o \in \mathcal{O}'} p^*(o) \cdot \frac{\omega(o)}{\sum_{o_j \in C(o)} \omega(o_j)} \tag{2}$$

Proof Consider a solution $[p_0, p_1, \ldots, p_m]^T$ of $LC(S, \mathcal{A})$. Then $\mathscr{P}_T(S') = \sum_{w_i \in \mathcal{W} \text{ s.t. } w_i \not\models_T S'} p_i$. Recalling the definition of *totally unexplained sequence*,

we can write

$$\mathscr{P}_T(S') = \sum_{w_i \in \mathcal{W} \text{ s.t. } w_i \not\models_T S'} p_i = \sum_{w_i \in \mathcal{W} \text{ s.t. } \forall a_i \in S', \nexists o \in w_i, a_i \in o} p_i \tag{3}$$

Note that the condition $\forall a_i \in S', \nexists o \in w_i, a_i \in o$ is satisfied by all worlds except those containing at least one occurrence intersecting S'. Therefore,

$$\sum_{w_i \in \mathcal{W} \text{ s.t. } \forall a_i \in S', \nexists o \in w_i, a_i \in o} p_i = 1 - \sum_{w_i \in \mathcal{W} \text{ s.t. } \exists o \in w_i, o \cap S' \neq \emptyset} p_i \tag{4}$$

Lower bound. Recalling that $\mathcal{O}^*$ is the set of occurrences intersecting S', and considering that the condition $\exists o \in w_i, o \cap S' \neq \emptyset$ is satisfied by all worlds w_i containing an occurrence $o \in \mathcal{O}^*$, with some worlds containing multiple such occurrences, we can write

$$\sum_{w_i \in \mathcal{W} \text{ s.t. } \exists o \in w_i, o \cap S' \neq \emptyset} p_i \leq \min\left\{1, \sum_{o \in \mathcal{O}^*} \sum_{w_i \in \mathcal{W} \text{ s.t. } \exists o \in w_i} p_i\right\} \tag{5}$$

Note that the argument $\sum_{w_i \in \mathcal{W} \text{ s.t. } \exists o \in w_i} p_i$ of the outer summation is the left-hand side of the constraint for occurrence o in the set of linear constraints of Definition 2. Therefore, combining Definition 2 and Equations 3, 4, and 5, we can write

$$\mathcal{P}_T(S') \geq 1 - \min\left\{1, \sum_{o \in \mathcal{O}^*} p^*(o) \cdot \frac{\omega(o)}{\sum_{o_j \in C(o)} \omega(o_j)}\right\} \qquad (6)$$

Upper bound. Consider a maximal intersecting set $\mathcal{O}' \in \mathcal{M}(\mathcal{O}^*)$. For any two occurrences $o_i, o_j \in \mathcal{O}'$, the sets of worlds $W_i = \{w \in \mathcal{W} | o_i \in w\}$ and $W_j = \{w \in \mathcal{W} | o_j \in w\}$ are disjoint, as $o_i \not\sim o_j$. Additionally, the condition $\exists o \in w_i, o \cap S' \neq \emptyset$ is satisfied in at least all worlds w_i containing an occurrence $o \in \mathcal{O}'$, therefore,

$$\sum_{w_i \in \mathcal{W} \text{ s.t. } \exists o \in w_i, o \cap S' \neq \emptyset} p_i \geq \sum_{o \in \mathcal{O}'} \sum_{w_i \in \mathcal{W} \text{ s.t. } o \in w_i} p_i \qquad (7)$$

As the above property holds for all $\mathcal{O}' \in \mathcal{M}(\mathcal{O}^*)$, we can conclude that

$$\sum_{w_i \in \mathcal{W} \text{ s.t. } \exists o \in w_i, o \cap S' \neq \emptyset} p_i \geq \max_{\mathcal{O}' \in \mathcal{M}(\mathcal{O}^*)} \sum_{o \in \mathcal{O}'} \sum_{w_i \in \mathcal{W} \text{ s.t. } o \in w_i} p_i \qquad (8)$$

Finally, combining Definition 2 and Equations 3, 4, and 8, we can write

$$\mathcal{P}_T(S') \leq 1 - \max_{\mathcal{O}' \in \mathcal{M}(\mathcal{O}^*)} \sum_{o \in \mathcal{O}'} p^*(o) \cdot \frac{\omega(o)}{\sum_{o_j \in C(o)} \omega(o_j)} \qquad (9)$$

Example 7 Consider again the observation sequence and occurrences of Fig. 3. We want to find upper and lower bounds for the probability $\mathcal{P}_T(S')$ that the sequence $S' = a_6, a_7, a_8, a_9, a_{10}$ is totally unexplained. For this example, $\mathcal{O}^* = \{o_3, o_4, o_5, o_6\}$ and $\mathcal{M}(\mathcal{O}^*) = \{\{o_3, o_4\}, \{o_4, o_5, o_6\}\}$. Applying Theorem 1 we obtain $\mathcal{P}_T(S') \geq 1 - 0.74 = 0.26$ and $\mathcal{P}_T(S') \leq 1 - \max\{0.36, 0.58\} = 0.42$. Note that, in this case, these bounds coincide exactly with the probability interval obtained by solving the maximization and minimization problems.

A consequence of Proposition 1 and Theorem 1 is the following theorem, which provides a sufficient condition for an observation not to be included in any unexplained behavior.

Theorem 2 *Let $I = S, \mathcal{A}, \tau, L$ be a UBP instance. Given an observation $a \in S$, if $1 - \sum_{o \in \mathcal{O} \text{ s.t. } a \in o} p^*(o) \cdot \frac{\omega(o)}{\sum_{o_j \in C(o)} \omega(o_j)} < \tau$, then there does not exist a subsequence S' of S s.t. (i) $a \in S'$, (ii) $\mathcal{P}_T(S') \geq \tau$, and (iii) $|S'| \geq L$.*

Proof Consider the sequence $S' = \langle a \rangle$. Then, $\mathcal{O}^* = \{o \in \mathcal{O} | o \cap S' \neq \emptyset\} = \{o \in \mathcal{O} | a \in o\}$. Therefore, $\mathcal{M}(\mathcal{O}^*) = \{\mathcal{O}^*\}$, i.e., there is only one maximal intersecting set in $\mathcal{O}^*$, and it coincides with $\mathcal{O}^*$. Applying Theorem 1, we can conclude that

$$\mathscr{P}_T\left(S'\right) \le 1 - \sum_{o \in \mathcal{O}^*} p^*(o) \cdot \frac{\omega(o)}{\sum_{o_j \in C(o)} \omega\left(o_j\right)} < \tau,$$

Now, consider a sequence $S'' \subseteq S$ s.t. $a \in S''$. Since $S' \subseteq S''$, from Proposition 1 we can conclude that

$$\mathscr{P}_T\left(S''\right) \le \mathscr{P}_T\left(S'\right) < \tau.$$

If the condition stated in the theorem above holds for an observation a, then we say that a is *sufficiently explained*. Note that checking whether an observation a is sufficiently explained does not require that we solve a set of linear constraints, since this can be done by simply summing the weighted probabilities of the occurrences containing a. Thus, this result yields a further efficiency. If a is sufficiently explained, then it can be disregarded for the purpose of identifying unexplained behaviors.

Given a UBP instance $I = \langle S, \mathscr{A}, \tau, L \rangle$ and a contiguous subsequence S' of S, we say that S' is a *candidate* iff (1) $|S'| \ge L$, (2) $\forall a \in S'$, a is not sufficiently explained, and (3) S' is maximal (i.e., there does not exist $S'' \ne S'$ is a subsequence of S'' and S'' satisfies (1) and (2)). We use *candidates*(I) to denote the set of candidate subsequences. If we look for totally unexplained behaviors that are contiguous subsequences of S, then Theorem 2 entails that candidate subsequences can be individually considered because there is no (contiguous) totally unexplained behavior spanning two different candidate subsequences.

3.2 Partially Unexplained Behaviors

We now present similar results, in terms of probability bounds, for partially unexplained behaviors, and show that lower and upper bounds for $\mathscr{P}_P(S')$ can be found without solving the optimization problem of Definition 4. In order to do so, we introduce the following preliminary definition.

Definition 7 (Non-conflicting sequence cover) Let O^* be a set of occurrences, and S' an observation sequence. A *non-conflicting sequence cover* of S' in O^* is a subset O' of O^* such that:

- $\forall o_i, o_j \in \mathcal{O}', o_i$ and o_j do not conflict; and
- $\forall a \in S', \exists o \in \mathcal{O}', a \in o$.

We use $\mathscr{C}(\mathcal{O}^*, S')$ to denote the set of all minimal sequence covers of S' in $\mathcal{O}^*$. Intuitively, a *non-conflicting sequence cover* of S' in $\mathcal{O}^*$ is a subset of non-conflicting occurrences in $\mathcal{O}^*$ covering[11] S'.

Theorem 3 *Consider a subsequence S' of an observation sequence S and the set $\mathcal{O}$ of occurrences identified in S, and let $\mathcal{O}^*$ be the set of occurrences $o \in \mathcal{O}$ such that $o \cap S' \neq \emptyset$. Then*

$$\mathscr{P}_P\left(S'\right) \geq 1 - \sum_{\mathcal{O}' \in \mathscr{C}(\mathcal{O}^*, S')} \min_{o \in \mathcal{O}'} p^*(o) \frac{\omega(o)}{\sum_{o_j \in C(o)} \omega(o_j)} \qquad (10)$$

$$\mathscr{P}_P\left(S'\right) \leq 1 - \sum_{o \in \mathcal{O}^* s.t. S' \subseteq o} p^*(o) \cdot \frac{\omega(o)}{\sum_{o_j \in C(o)} \omega(o_j)} \qquad (11)$$

Proof Consider a solution $[p_0, p_1, \ldots, p_m]^T$ of $LC(S, \mathscr{A})$. Then $\mathscr{P}_P(S') = \sum_{w_i \in \mathscr{W} s.t. w_i \not\models_P S'} p_i$. Recalling the definition of *partially unexplained sequence*, we can write

$$\mathscr{P}_T\left(S'\right) = \sum_{w_i \in \mathscr{W} s.t. w_i \not\models_T S'} p_i = \sum_{w_i \in \mathscr{W} s.t. \exists a_i \in S', \nexists o \in w_i, a_i \in o} p_i \qquad (12)$$

Note that the condition $\exists a_i \in S', \nexists o \in w_i, a_i \in o$ is satisfied by all worlds except those where each observation $a_i \in S'$ is part of an occurrence, that is the sequence is *totally explained* in those worlds. Therefore,

$$\sum_{w_i \in \mathscr{W} s.t. \exists a_i \in S', \nexists o \in w_i, a_i \in o} p_i = 1 - \sum_{w_i \in \mathscr{W} s.t. \forall a_i \in S', \exists o \in w_i, a_i \in o} p_i \qquad (13)$$

Lower bound. Given any two non-conflicting sequence covers $\mathcal{O}'$ and $\mathcal{O}''$, the sets of worlds $\mathscr{W}' = \{w \in \mathscr{W} | \mathcal{O}' \subseteq w\}$ and $\mathscr{W}'' = \{w \in \mathscr{W} | \mathcal{O}'' \subseteq w\}$ are disjoint, as at least one occurrence in $\mathcal{O}'$ conflicts with at least one occurrence in $\mathcal{O}''$. Additionally, the condition $\forall a_i \in S', \exists o \in w_i, a_i \in o$ is satisfied by all worlds w_i containing a non-conflicting cover of S'. Thus, we can write

$$\sum_{w_i \in \mathscr{W} s.t. \forall a_i \in S', \exists o \in w_i, a_i \in o} p_i = \sum_{\mathcal{O}' \in \mathscr{C}(\mathcal{O}^*, S')} \sum_{w_i \in \mathscr{W} s.t. \mathcal{O}' \subseteq w_i} p_i \qquad (14)$$

---

[11] This is a variant of the set cover problem. This is known to be NP-complete, however we need to solve only small instances of this problem, so complexity is not an issue.

Consider any non-conflicting sequence cover $\mathcal{O}' \in \mathscr{C}(\mathcal{O}^*, S')$. The set $\mathscr{W}' = \{w_i \in \mathscr{W} | \mathcal{O}' \subseteq w_i\}$ of worlds containing all the occurrences in $\mathcal{O}'$ is a subset of the set of worlds containing o, for each $o \in O'$. Therefore,

$$\sum_{w_i \in \mathscr{W} \, s.t. \, \mathcal{O}' \subseteq w_i} p_i \leq \min_{o \in \mathcal{O}'} \sum_{w_i \in \mathscr{W} \, s.t. \, o \in w_i} p_i \tag{15}$$

Finally, considering that the above property holds for any $\mathcal{O}' \in \mathscr{C}(\mathcal{O}^*, S')$, and combining Definition 2 and Equations 12, 13, 14, and 15, we can write

$$\mathscr{P}_P(S') \geq 1 - \sum_{\mathcal{O}' \in \mathscr{C}(\mathcal{O}^*, S')} \min_{o \in \mathcal{O}'} p^*(o) \frac{\omega(o)}{\sum_{o_j \in C(o)} \omega(o_j)}$$

Upper bound. Consider the set $\mathcal{O}' = \{o \in \mathcal{O}^* | S' \subseteq o\}$ of all the occurrences o that cover S', i.e., the occurrences such that $\{o\}$ is a sequence cover for S' in $\mathcal{O}^*$. S' is *totally explained* in at least all the worlds containing any of the occurrences in $\mathcal{O}'$. Note that any two set of worlds W_i and W_j containing $o_i \in \mathcal{O}'$ and $o_j \in \mathcal{O}'$, respectively are disjoint, as $o_i \nsim o_j$. Therefore,

$$\sum_{w_i \in \mathscr{W} \, s.t. \, \forall a_i \in S', \exists o \in w_i, a_i \in o} p_i \geq \sum_{o \in \mathcal{O}^* \, s.t. \, S' \subseteq o} \sum_{w_i \in \mathscr{W} \, s.t. \, o \in w_i} p_i \tag{16}$$

Finally, combining Definition 2 and Equations 12, 13, and 16, we can write

$$\mathscr{P}_P(S') \leq 1 - \sum_{o \in \mathcal{O}^* \, s.t. \, S' \subseteq o} p^*(o) \cdot \frac{\omega(o)}{\sum_{o_j \in C(o)} \omega(o_j)}$$

Example 8 Consider again the observation sequence and occurrences of Fig. 3. We want to find upper and lower bounds for the probability $\mathscr{P}_P(S'')$ that the sequence $S'' = a_7, a_8$ is partially unexplained. For this example, $\mathscr{C}(\mathcal{O}^*, S'') = \{\{o_3, o_5\}, \{o_4\}\}$. Applying Theorem 3 we obtain $\mathscr{P}_P(S'') \geq 1 - 0.36 = 0.64$ and $\mathscr{P}_P(S'') \leq 1 - 0.2 = 0.8$. Note that, in this case, these bounds coincide exactly with the probability interval obtained by solving the maximization and minimization problems.

4 Algorithms

Even though our framework can assess the probability that an arbitrary subsequence of S is unexplained, we propose algorithms that search for contiguous subsequences of S, as we believe that contiguous subsequence can more easily be interpreted by users (nevertheless, the algorithms could be easily modified to identify also non-contiguous unexplained subsequences).

We now present algorithms to find totally and partially unexplained behaviors. These algorithms are a variant of the algorithms in [14]: while the algorithms in [14] compute the *exact* probability that a sequence is unexplained, the algorithms in this chapter compute an *approximate* probability that a sequence is unexplained by lever-aging the properties shown in Sect. 3.

Given an observation sequence $S = \langle a_1, \ldots, a_n \rangle$, we use $S(i,j)$ $(1 \leq i \leq j \leq n)$ to denote the subsequence $S = a_i, \ldots, a_j$.

The FindTUB algorithm computes totally unexplained behaviors in an observation sequence S. Leveraging Theorem 2, FindTUB only considers candidate subsequences of S. When the algorithm finds a sequence $S'(start, end)$ of length at least L having a probability of being unexplained greater than or equal to τ (line 6), then the algorithm makes it maximal by adding observations on the right. Instead of adding one observation at a time, $S'(start, end)$ is extended of L observations at a time until its probability drops below τ (lines 8–13); then, the exact maximum length of the unexplained behavior is found by performing a binary search between s and e (line 16). Note that $\mathscr{P}_T$ is estimated by applying Theorem 1.

The FindPUB algorithm computes all partially unexplained behaviors. To find an unexplained behavior, it starts with a sequence of a certain length (at least L) and adds observations on the right of the sequence until its probability of being unexplained is greater than or equal to τ. As in the case of FindTUB, this is done not by adding one observation at a time, but adding L observations at a time (lines 7–11) and then determining the exact minimal length by performing a binary search between s and e (line 16). The sequence is then shortened on the left making it minimal by performing a binary search instead of proceeding one observation at a time (line 23). Note that $\mathscr{P}_P$ is estimated by leveraging Theorem 3.

5 Experimental Results

In this section, we present the results of experiments we conducted on a prototype implementation of the proposed framework. We evaluated running time of the algorithms as well as accuracy. In the following, we first describe the experimental setup (Sect. 5.1), and then report the results on the scalability (Sect. 5.2) and accuracy (Sect. 5.3) of our framework.

Algorithm 1 FindTUB(I)

Require: UBP instance $I = \langle S, \mathcal{A}, \tau, L \rangle$
Ensure: Set $O^{tu}(I)$ of totally unexplained behaviors
1: $O^{tu}(I) = \emptyset$
2: **for all** $S' \in relevant(I)$ **do**
3:     $start = 1$
4:     $end = L$
5:     **repeat**
6:        **if** $\mathcal{P}_T\big(S'(start, end)\big) \geq \tau$ **then**
7:           $end' = end$
8:           **while** $end < |S'|$ **do**
9:               $end = \min\{end + L, |S'|\}$
10:              **if** $\mathcal{P}_T\big(S'(start, end)\big) < \tau$ **then**
11:                 **break**
12:              **end if**
13:           **end while**
14:           $s = \max\{end - L, end'\}$
15:           $e = end$
16:           $end = \max\{mid \in \mathbb{N}^+ \mid s \leq mid \leq e \wedge \mathcal{P}_T\big(S(start, mid)\big) \geq \tau\}$
17:           $S'' = S'(start, end)$
18:           Add S'' to $O^{tu}(I)$
19:           $start = start + 1$
20:           $end = start + |S'| - 1$
21:        **else**
22:           $start = start + 1$
23:           $end = \max\{end, start + L - 1\}$
24:        **end if**
25:     **until** $end > |S'|$
26: **end for**
27: **return** $O^{tu}(I)$

5.1 Experimental Setup

All experiments were conducted on a dataset consisting of network traffic captured over a 24-h period from the internal network of an academic institution. We used (1) Wireshark (http://www.wireshark.org/) to capture network traffic and generate the sequence of packets, and (2) Snort (http://www.snort.org) to analyze such traffic and generate the sequence of alerts.

Algorithm 2 FindPUB(I)

Require: UBP instance $I = \langle S, \mathcal{A}, \tau, L \rangle$
Ensure: Set $O^{pu}(I)$ of partially unexplained behaviors
1: $O^{pu}(I) = \emptyset$
2: $start = 1$
3: $end = L$
4: **while** $end \leq |S|$ **do**
5:     **if** $\mathcal{P}_P\big(S(start, end)\big) < \tau$ **then**
6:         $end' = end$
7:         **while** $end \leq |S|$ **do**
8:             $end = \min\{end + L, |S|\}$
9:             **if** $\mathcal{P}_P\big(S(start, end)\big) \geq \tau$ **then**
10:                **break**
11:             **end if**
12:         **end while**
13:         **if** $\mathcal{P}_P\big(S(start, end)\big) \geq \tau$ **then**
14:             $s = \max\{end' + 1, end - L + 1\}$
15:             $e = end$
16:             $end = \min\{mid \in \mathbb{N}^+ | s \leq mid \leq e \wedge \mathcal{P}_P\big(S(start, mid)\big) \geq \tau\}$
17:         **else**
18:             **return** $O^{pu}(I)$
19:         **end if**
20:     **end if**
21:     $s' = start$
22:     $e' = end - L + 1$
23:     $start = \max\{mid \in \mathbb{N}^+ | s' \leq mid \leq e' \wedge \mathcal{P}_P\big(S(mid, end)\big) \geq \tau\}$
24:     $S' = S(start, end)$
25:     Add S' to $O^{pu}(I)$
26:     $start = start + 1$
27:     $end = start + |S'| - 1$
28: **end while**
29: **return** $O^{pu}(I)$

Figure 5 illustrates the experimental setup. As the number of alerts returned by the IDS may be relatively high, the *Alert Aggregation* module, that takes as input the identified alerts, can optionally aggregate multiple alerts triggered by the same event into a macro-alert, based on a set of ad hoc aggregation rules. For instance, we defined rules to aggregate alerts such that protocol, source address, and destination address of suspicious traffic are the same, and the alerts are within a given temporal window. In other words, the events triggering such alerts will be treated as a single event, thus reducing the amount of data to be processed.

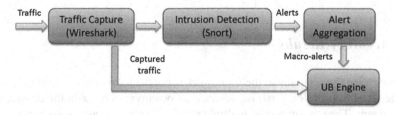

Fig. 5 Experimental setup

Fig. 6 Execution time
($\tau = 0.6$, $L = 10$, 50, 75,
100)

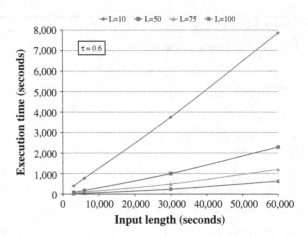

5.2 Scalability Results

We measured the running time of FindTUB for different values of τ and L,
varying the length of the data stream to be analyzed. More specifically, in one case
we set the threshold τ to 0.6 and used different values of L, namely 10, 50, 75, 100.
In the other case, we fixed the value of L to 50 and varied the threshold τ giving it
the values 0.4, 0.6, 0.8.

Figure 6 shows the processing time of FindTUB as a function of the data
stream length (expressed in seconds) for different values of L. Not surprisingly, the
running time increases as the input stream size grows. Moreover, the processing
time decreases as the value of L increases because Algorithm FindTUB can move
forward in the data stream more quickly for higher values of L. Notice also that the
running times is much lower when $L \geq 50$.

Figure 7 shows the processing time of FindTUB as a function of the data
stream length for different values of τ. Also in this case the running time gets
higher as the input stream length increases. The running time is lower for higher
values of τ because the pruning strategy of Algorithm FindTUB becomes more
effective with higher threshold values. Moreover, in this case, the running time
becomes much lower when $\tau \geq 0.6$.

Both figures show that our algorithm scales well—notice that the running time
linearly grows w.r.t. the length of the input.

5.3 Accuracy Results

We measured the accuracy of the framework using the following procedure. Let $\mathscr{A}$
be the set of Snort rules. First, we detected all occurrences of $\mathscr{A}$ in the considered
data stream. Then, we executed multiple runs of FindTUB, and at each run i, we

Fig. 7 Execution time
($L = 50$, $\tau = 0.4, 0.6, 0.8$)

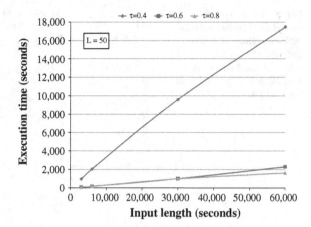

ignored a different subset $\mathscr{A}_i$ of $\mathscr{A}$. Clearly, ignoring models in $\mathscr{A}_i$ is equivalent to not having those models available. Thus, occurrences of ignored behaviors are expected to have a relatively high probability of being unexplained as there is no model for them. We measured the fraction of such occurrences that have been flagged as unexplained by FindTUB for different values of τ, namely 0.4, 0.6, 0.8 (L was set to 50).

We considered two settings: one where only *ICMP rules* in $\mathscr{A}$ were ignored, and another one where only *preprocessor rules* in $\mathscr{A}$ were ignored. The average accuracy in the former and latter case is shown in Tables 2, 3, respectively. Notice that the accuracy decreases as the threshold value increases since higher thresholds are more restrictive conditions for a sequence to be unexplained. Notice also that in both cases there is no difference in accuracy between $\tau = 0.6$ and $\tau = 0.8$; this is because, in this case, the same unexplained sequences were found. These results show that our framework achieved high accuracy.

We also evaluated the effect of the threshold value on the percentage of unexplained traffic. We considered three different settings, each characterized by a

Table 2 Accuracy when ICMP rules are ignored

τ	Accuracy (%)
0.4	95.10
0.6	78.75
0.8	78.75

Table 3 Accuracy when preprocessor rules are ignored

τ	Accuracy (%)
0.4	84.21
0.6	73.68
0.8	73.68

Fig. 8 Percentage of
unexplained traffic vs. τ for
different values of L and
different sets of rules.
a Unexplained traffic for
$L = 50$. **b** Unexplained traffic
for $L = 100$

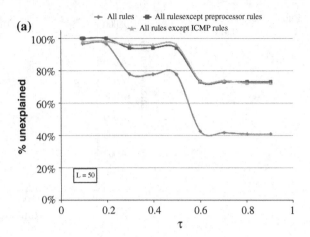

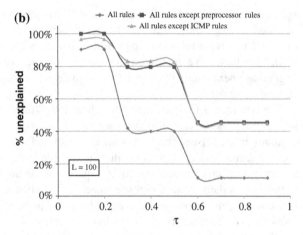

different set of available rules (all Snort rules are available, all Snort rules except
for preprocessor rules are available, and all rules but ICMP rules are considered),
and measured the percentage of unexplained traffic as the threshold varies from 0.1
to 1. We carried out the experiments for two different values of L, namely $L = 50$
and $L = 100$—the results are reported in Fig. 8a, b, respectively. Both figures
show that by disregarding rules the percentage of unexplained traffic increases, but
there is no substantial difference between disregarding preprocessor rules and
disregarding ICMP rules. Furthermore, the percentage of unexplained traffic
decreases as the threshold increase because higher threshold values impose more
restrictive conditions for a sequence to be unexplained. Finally, the results for
$L = 100$ show lower percentages of unexplained traffic than the case $L = 50$ as
$L = 100$ is a more restrictive condition for a sequence to be unexplained and thus
the unexplained traffic is expected to be less in this case. This trend is also
confirmed by the results of Fig. 9 where we show how the percentage of unexplained traffic varies as the threshold value goes from 0.1 to 1 and different values

Fig. 9 Percentage of unexplained traffic vs. τ for different values of L (all IDS rules are used)

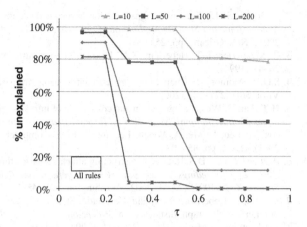

of L are considered. In this case L was set to 10, 50, 100, 200 and all IDS rules were considered.

6 Conclusions

In this chapter, we presented a probabilistic framework to identify unexplained behavior in network intrusions. Intrusion detection and alert correlation methods rely on models encoding a priori knowledge of either normal or malicious behavior, but are incapable of quantifying how well the underlying models explain what is observed on the network. Our framework addresses this limitation, by evaluating the probability that a sequence of events is unexplained, given a set of models. We derived some important properties of the framework that can be leveraged to estimate this probability efficiently. The proposed framework can operate both at the intrusion detection level and at the alert correlation level. Experimental results show that the algorithms are accurate and scale linearly with the size of the observation sequence. This confirms the validity of our approach and motivates further research in this direction.

Acknowledgments The work presented in this chapter is supported in part by the Army Research Office under MURI award number W911NF-09-1-05250525, and by the Office of Naval Research under award number N00014-12-1-0461. Part of the work was performed while Sushil Jajodia was a Visiting Researcher at the US Army Research Laboratory.

References

1. P. García-Teodoro, J. Díaz-Verdejo, G. Maciá-Fernández, E. Vázquez, Anomaly-based network intrusion detection: techniques, systems and challenges. Comput. Secur. **28**(1–2), 18–28 (2009)

2. A. Jones, S. Li, Temporal signatures for intrusion detection, in *Proceedings of the 17th Annual Computer Security Applications Conference (ACSAC 2001)* (IEEE Computer Society, 2001), New Orleans, pp. 252–261

3. B. Mukherjee, L.T. Heberlein, K.N. Levitt, Network intrusion detection. *IEEE Netw.* **8**(3), 26–41 (1994)

4. S.O. Al-Mamory, H. Zhang, Ids alerts correlation using grammar-based approach. J. Comput. Virol. **5**(4), 271–282 (2009)

5. H. Debar, A. Wespi, Aggregation and correlation of intrusion-detection alerts, in *Proceedings of the 4th International Symposium on Recent Advances in Intrusion Detection (RAID 2001)*, eds. W. Lee, L. Mé, A. Wespi. Lecture Notes in Computer Science, vol. 2212 (Springer, 2001), Davis, pp. 85–103

6. P. Ning, Y. Cui, D.S. Reeves, Constructing attack scenarios through correlation of in- trusion alerts, in *Proceedings of the 9th ACM Conference on Computer and Communications Security(CCS 2002)* (ACM, 2002), Washington, pp. 245–254

7. S. Noel, E. Robertson, S. Jajodia, Correlating intrusion events and building attack scenarios through attack graph distances, in *Proceedings of the 20th Annual Computer Security Applications Conference (ACSAC 2004)* (2004), Tucson, pp. 350–359

8. L. Wang, A. Liu, S. Jajodia, Using attack graphs for correlating, hypothesizing, and predicting intrusion alerts. Comput. Commun. **29**(15), 2917–2933 (2006)

9. J.P. Anderson, Computer security threat monitoring and surveillance. Technical report, James Anderson Co., Fort Washington, Apr 1980

10. O. Sheyner, J. Haines, S. Jha, R. Lippmann, J.M. Wing, Automated generation and analysis of attack graphs, in *Proceedings of the 2002 IEEE Symposium on Security and Privacy (S&P 2002)*, Berkeley, 2002, pp. 273–284

11. X. Qin, A probabilistic-based framework for INFOSEC alert correlation. Ph.D. thesis, Georgia Institute of Technology, 2005

12. X. Qin, W. Lee, Statistical causality analysis of INFOSEC alert data, in *Proceedings of the 6th International Symposium on Re- cent Advances in Intrusion Detection (RAID 2003)*, eds. G. Vigna, C. Kruegel, E. Jonsson. Lecture Notes in Computer Science, vol. 2820 (Springer, 2003), Pittsburgh pp. 73–93

13. A.J. Oliner, A.V. Kulkarni, A. Aiken, Community epidemic detection using time- correlated anomalies, in *Proceedings of the 13th International Symposium on Recent Advances in Intrusion Detection (RAID 2010)*, eds. S. Jha, R. Sommer, C. Kreibich. Lecture Notes in Computer Science, vol. 6307 (Springer, 2010), Ottawa, pp. 360–381

14. M. Albanese, C. Molinaro, F. Persia, A. Picariello, V.S. Subrahmanian, Finding "un- explained" activities in video, in *Proceedings of the 22nd International Joint Conference on Artificial Intelligence (IJCAI 2011)*, Barcelona, 2011, pp. 1628–1634

15. M. Albanese, S. Jajodia, A. Pugliese, V.S. Subrahmanian, Scalable analysis of attack scenarios, in *Proceedings of the 16th European Symposium on Research in Computer Security (ESORICS 2011)* (Springer, 2011), Leuven, pp. 416–433

Applying Cognitive Memory to CyberSecurity

Bruce McCormick

1 Introduction

Too much time and effort is spent on analyzing crime after the fact with too little emphasis on detecting the crime in progress. Through the use of cognitive learning and recognition, this imbalance can be improved towards more effective and higher performance techniques.

This chapter deals with a hardware based non-linear classifier than learns from the data and thereby can be used to detect deviations from the norm, or recognize patterns in data. In essence, the memory itself learns patterns and reacts to an input in parallel. It finds the closest match in a fixed amount of time regardless of the number of comparisons. Neural networks and nonlinear classifiers (more general than linear) are known to be good pattern recognition engines with the ability to generalize using imprecise or incomplete data. This chapter is about a commercially available neural network that is natively implemented in a hardware component comprising of 1024–256 bytes each—"cognitive memories" for pattern learning and recognition performance/watt acceleration. Additionally this hardware is agnostic to the type of digital data making it useful for biometric pattern matching, fuzzy matching, extracting profiles, and supporting behavioral analysis. Exact matching analysis like hash function comparison of 1 versus large N of unordered, non-indexed files, is also supported.

What this hardware engine is good for is for sorting, clustering, learning, recognizing and providing anomaly detection in data patterns that are presented to it in parallel. First is an explanation of the hardware architecture that performs these functions.

B. McCormick (✉)
CogniMem Technologies, Folsom, CA, USA
e-mail: bruce.mccormick@cognimem.com

R. E. Pino (ed.), *Network Science and Cybersecurity*,
Advances in Information Security 55, DOI: 10.1007/978-1-4614-7597-2_4,
© Springer Science+Business Media New York 2014

2 Architectural Overview

The architecture shown below in Fig. 1 illustrates a memory based parallel processing unit wherein the input is broadcast to all stored "prototypes" (or "models", "data vectors") simultaneously as an associative memory. Each prototype does an L1 (or sum of absolute differences or Manhattan distance) or Lsup calculation of what is stored in that memory location versus what is broadcast at the input. The results are then used to determine how close the input prototype is to the "learned" or stored information. When this information is close enough in match—it is categorized with that specific neuron.

In kNN (k Nearest Neighbor) mode—each calculated distance is compared to all others simultaneously (in a constant 16 clock cycles) to determine who is the closest to the input pattern. This results in one neuron output being available—the information communicated is the distance and category associated with that neuron. Subsequent reads can find the next "k" closest neighbors.

In Radial Basis Function (RBF/RCE) mode it is helpful to think of learning and recognition. During recognition, the component uses the min/max "filter" on the output distance to determine which and how many neuron categories the input is associated with. There can be a match (1 category), no match (0 categories) or uncertain (where more than one category fires). In this event, it requires further analysis or features to be looked at to make a decision. In the RBF/RCE learning mode, a prototype is stored in the first neuron location representing a data pattern

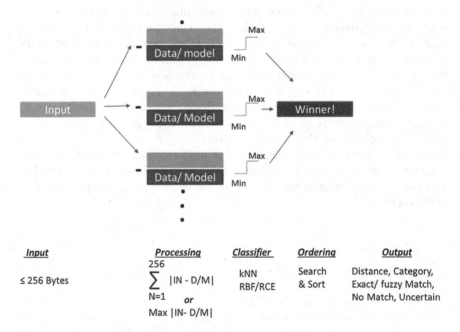

Input	Processing	Classifier	Ordering	Output
≤ 256 Bytes	$\sum_{N=1}^{256} |IN - D/M|$ or $Max |IN - D/M|$	kNN RBF/RCE	Search & Sort	Distance, Category, Exact/ fuzzy Match, No Match, Uncertain

Fig. 1 Architecture of Cognitive Memory

to be learned. When the next pattern comes, it is looked at to see if it is close enough to the initial stored pattern to be recognized (using the min/max filter)—if not, then another neuron/prototype is made with the same category. Now 2 neurons are committed with different patterns that are used to identify the inputs for the same category. If the new prototype submitted is a different object or piece of data with a different category—the hardware will make the determination if one of the previously stored prototypes will claim it is a match of another category. If so, then the hardware automatically adjusts the min/max filter used on the distance information to exclude the new data point from being recognized by the other neurons. This is the learning process called "RCE" or Restricted Coloumb Energy. Once all training examples have been trained in this way, one can use a set of validation feature sets to verify the accuracy of the classifier. When a previously unseen data input is submitted, the classifier will either recognize (match within the limits) not match (no neurons fire) or have neurons fire from different categories (uncertain) requiring further processing. One can have false positives—which can be mitigated with conservative learning (models have to be closer to the stored models to agree), or false negatives—which require more prototypes to be searched against to mitigate (Fig. 1).

This type of memory based architecture eliminates the instruction execution cycle in traditional processing and the movement of data that causes the bottleneck between CPU and memory. This is achieved by having the memory itself doing the pattern recognition work in parallel in response to an input or "stimulus". This architecture is quite useful for many applications in the internet framework (Fig. 2).. Examples include validating a sender or receiver of a file with biometrics (offline or real time), detecting a virus or denial of service attack- looking at

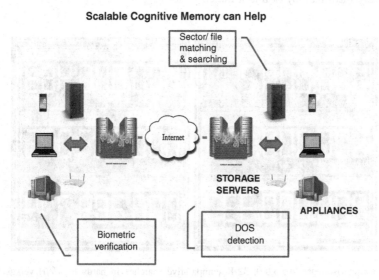

Fig. 2 Using Cognitive Memory in the Internet Framework

packets/headers for known malicious signatures or searching through files, data for suspicious activities or correlation to patterns being searched for.

The Cognimem architecture can also be used to learn (associate) historical variables to outcomes, then apply this to present data. One can also think of it as an associative memory where each chip maps 2(2048) in a sparse matrix to 2(15) categories.

3 Biometric Example

In this relatively straightforward example, one would only need to pack the data into 256 bytes. This data is stored in the cognitive memory in an unordered, non-indexed fashion. If the data is a fingerprint for example, each fingerprint minutia is stored as a prototype into the cognitive memory for all the input patterns to be compared against. With this architecture, virtually an unlimited number of prints can be compared against in a fixed amount of time regardless of the number of prints. This comparison of 1–256 byte feature set against N other feature sets is done in 10 µs. This is significant when one is looking for a "needle in a haystack" quickly and doesn't have the time for the computationally intensive comparisons in a sequential fashion. This same technique can be used whether it is an iris signature, voiceprint, or other unique identifiers of an individual to determine identification or access rights. If confidence needs to be increased, this can be combined with other identifiers to gain assurance that the match is a high probability event. In Fig. 3 a comparison is made of doing 320 versus 3200 versus 32,000 finger prints being searched against (time includes system overhead for displaying information) to find a match.

Fig. 3 Comparison of 320, 3200, 32 K comparative searches in hardware (*left*) versus S/W (*right*)

First the recognition was performed in simulation mode and then was performed through the 40 k neuron hardware system. As can be observed—data base 1, 2 and 3 took 0.46, 7.07 and 284.7 ms for finding the best match in simulation mode.

Conversely, the 40 K neuron Cognimem hardware system took 0.27, 0.38 and 0.337 ms for each data base (@ less than 7 W typical) respectively. Figure 3 is a screen shot for the fingerprint recognition performance using a 40 k neuron system. (Difference in times shown is due to host overhead—comparison processing is the same) Thus an unknown biometric print can be checked against a large data base in a fixed amount of time. There are many algorithms in use for doing feature extraction of an image, object, scene etc., some examples include SIFT, SURF, ORB and Freak. A compute intensive process then presents itself as these features are then compared against a known data base.

An example could be detecting an airplane for identification or cars from their logos. For fastest operation, the Cognimem memory would need to be sufficiently large to store all the known feature vectors (assuming each vector is less than or equal to 256 bytes). When a new object presents itself—the feature extractor submits it to the Cognimem device and in 10 μs the closest one is found.

For some applications this may be faster than what the constraints are and cost could be traded off against speed. One can do the reverse and store the incoming vectors into the cognitive memory and then have the data base compared against it iteratively to find the closest match.

4 Clustering of Data Example

One can take data never seen before and cluster it based on desired comparison examples or "random" start points for association. Figure 4 below is an example approach to this:

If you assume a hypercube (hyperspectral image for example) of data that is 1 mpixel by 1 mpixel and is made of 256 planes-each 1 byte deep (256 bytes per pixel), the process is as follows for clustering:

(1) Start with the first pixel and learn it in a neuron. Set a minimum influence field and a maximum influence field for that neuron (shown by solid diamond and dotted diamond respectively). This is your start point for clustering.
(2) Go to the next pixel and determine L1 from the first vector. If it's distance is within the minif, you do not store the data in the neurons, but keep track of the vector being similar (same category) in an external counter. If it is between the min and max, go to 3, otherwise go to step 4.
(3) If the new pixel L1 is between the min and maxif of the first pixel, you want to also learn this pixel in this category to keep track of contiguous (still similar) vectors, but learn it to allow for migration in a "direction" of the similar data. You accomplish this on the Cognimem device by first learning this new pixel under context 0 (this shrinks the max influence field of the first neuron to

Fig. 4 Clustering of data

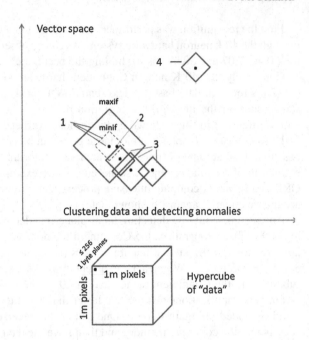

Clustering data and detecting anomalies

Hypercube
of "data"

exclude the new data point- not shown in diagram) and then learn the new pixel as another neuron—but same category—and so on. This is the same process used for tracking an object.

(4) If the new pixel is outside the previous neuron's influence field, then a new category (cluster) and neuron is created.

The new data points will either be classified as "known" (within the influence field of existing neurons) "unknown" (need a new category to learn it) or "uncertain" (Two different categories claim it is within their influence fields). If you have two clusters forming that eventually overlap each other—this will be flagged as "uncertain" where the supervisor can go into the data and see if they are the same category or not based on statistics criteria, other info, etc.

When you have completed the task—clusters of a single neuron or a few neurons (per your criteria) will be the "outliers".

5 Adaptive Tracking, Virus Detection

The most commonly used technique for virus detection is signature detection. This consists of taking the content or header of a file and comparing it against signatures in an anti-virus data base. The Cognimem device can certainly be used to compare incoming data against an unordered, non-indexed list of known malicious signatures. However, more sophisticated viruses are out there such as code obfuscating by re-ordering assembly instructions, inserting no-ops or dead code and replacing subsections of code that have an equivalent operation—thus "morphing" the signature. Tracking of information that morphs in such a way can be challenging—particularly if the virus detection can only detect exact matches. If a known "signature" or code string (represented digitally) is learned and is expected to change, then as it morphs, one can commit a new neuron before the first neuron loses the recognition. This can be accomplished by monitoring the distance that the new example is away from the originally learned example. As it becomes further away, new neurons can be committed by learning the new example and adding to the same category as the first example. By learning the new example, the number of neurons firing plus the distances can be tracked and used in the decision of identifying the morphing example. Another approach would be to train the Cognitive Memory virus detection system with metamorphic variants that are equivalent in function to the original code or signature. By using the fuzzy generalization properties of the nonlinear classifiers in the Cognimem device- unseen but similar variants are more likely to be caught and flagged for further inspection.

6 Hash Function

Many applications use hash functions to tag a file uniquely. This can be used to eliminate extra copies of a file (de-duplication), verify access rights or whether or not a file has been tampered with between sender and receiver among other things. Cognitive memory can be used in kNN mode to find exact matches to compare hash functions of an incoming file against a large list or to check to see if the file is one that this user is qualified to view or if there has been modifications.

  If data is streaming at high bandwidth, one would take the number of patterns (prototypes, models etc.) that are being searched for and having sufficient hardware with these patterns pipelined to handle the throughput being sought after. For instance @ 20 gbit ethernet and 2000 patterns—one would have 100 stacks of two components (each being able to store 1024 patterns) where the data flows into the parallel search engine in a round robin methodology so that the output would also stream at the same data rate with a fixed delay for the parallel pipelined search.

  By picking the right attributes in the data that are being searched, one could then be searching for behavioral markers—such as virus morphing from a known start point. If the file is sufficiently close to the original file that is a known malware—one could kick it out for further analysis before sending it on.

  Link analysis or breadth first search type operations can also be used to determine associations of information- who knew who, what sources are connected together.

7 Product Overview

The product behind these novel solutions is Cognimem's CM1 K comprising of 1024 cognitive memories or processing elements. A block diagram and die photo are shown in Figs. 5 and 6 respectively. Each processing element learns by storing digital data up to 256 bytes and compares these stored models to a broadcasted vector to be fuzzy or exact matched against. The learning process can be in situ real time, updated at a later date to include new knowledge or preloaded from previous training offline. For simple kNN, comparison vectors are just written into the processing memory elements prior to finding the closest neighbors.

This architecture is set up as a 3-layer network with an input layer, processing layer and an output search/sort layer, all done directly in hardware. Each silicon chip contains 1024 of the processing elements and can share the recognition, vector comparison task with N number of components on the same parallel bus.

A modular board ("module") incorporating 4 of these components that are connected through a dynamically reconfigurable Lattice XP2 FPGA was also built. This architecture provides the customer with the flexibility of creating user-defined interconnect topologies as required by application constraints. Local fast non-volatile Magnetoresistive Random Access Memory (MRAM) is also provided for fast local loads of pre-trained datasets during system power-up. Conversely, the

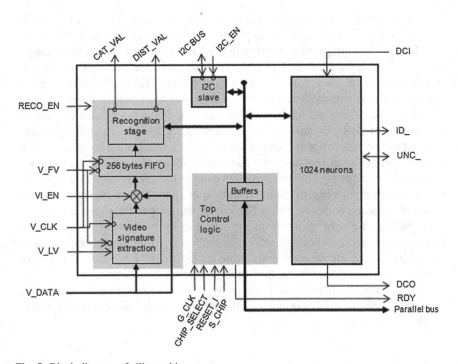

Fig. 5 Block diagram of silicon chip

Fig. 6 Die photo

same local storage can be used to store large datasets to be compared to a single pattern loaded into the system at runtime at real-time speeds (Figs. 7, 8).

A cardinal connect topology which allows arrays of modules as well as a vertical connection through a "spine" of up to 14 modules (Ten were used in this system) each is possible. This flexible architecture can be configured to have all

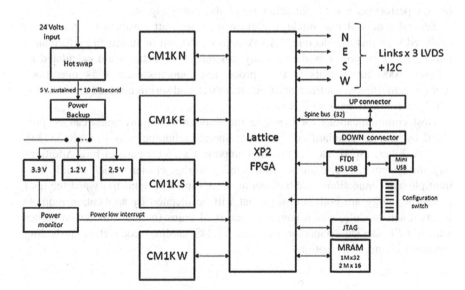

Fig. 7 Block diagram of modular board

Fig. 8 Top down picture of
4,096 processing element
board

processing elements working together on a common problem (like finding 1 iris or
fingerprint among 40 thousand with very low latency) or subsets of the same
problem (like searching for an anomaly in a hyperspectral image) where the image
is partitioned to be processed in parallel on different modules. The processing
element that finds the closest match alerts the user in as little as 10 μs regardless of
the number of data vectors being searched against, giving unprecedented low
latency performance for large data base applications (Fig. 9).

Several orders of magnitude of performance per watt advancement have been
achieved with this architecture. 0.13 Petaops equivalent of sustained performance
under 250 W of power is theoretically possible. 0.13 Petaops are calculated as
follows: 1000 components × 1024 processing elements each × 5+ operations
(compare, multiplex, subtract, accumulate, search and sort minus load and store) ×
256 byte connections × 100 K/sec.

Host communications for handling the throughput requirements can be provided by various standard computer peripheral technologies (e.g. USB 2.0/3.0,
eSATA, Fiber, iSCSI, etc....) via the component's 16-bit parallel bus. Additionally, the system can be segmented into smaller storage clusters which would allow
multiple bus connections, one bus feeding each cluster. Connectivity and the final
system topology are both decisions that will be dictated by application requirements. Each module, which contains four 1024 chips (memory modules) comes
with a USB device connector and four LVDS cardinal connectors, providing
countless connectivity options.

Fig. 9 Example column of
boards

Communications between he host and the system occur through register transfer
commands. These commands control the load/learning, comparison, and retrieval
functions of the system. There are also three SDKs available to help accelerate the
adoption of this technology, C, NET, and Java libraries.

For more information see www.cognimem.com, www.digikey.com

Understanding Cyber Warfare

Yan M. Yufik

> *"It doesn't matter how far ahead you see if you don't*
> *understand what you are looking at."*
> Garry Kasparov, 2007. How Life Imitates Chess. p. 50

The history of computing devices goes back to the invention of the abacus in Babylonia in the sixteenth century BC. In the three and a half millennia which followed, a variety of calculating devices were introduced, some of them stunningly sophisticated but all sharing a common limitation: a device could store data only as long as programs for the data manipulation remained in the mind of the user. The first breakthrough occurred circa 1830 AD, when Babbage designed an "Analytical Engine", run by programs stored on punch cards. The digital revolution was brought about by the idea that programs can be stored in the same medium as the data [1]. Finally, ARPANET was introduced in 1970, creating a foundation for the internet of today. Students of history readily point out that the benefits of technological advancement have often been accompanied by unintended and unforeseeable problems. The digital revolution and the emergence of the World Wide Web is the case in point: both brought about unprecedented benefits but also introduced a new form of warfare wrought with unprecedented risks and vulnerabilities. This article discusses the nature of such risks and vulnerabilities and the approaches to mitigating them. The discussion is preliminary, aimed at articulating suggestions for further research rather than detailing a particular method. The following two notions are addressed.

First, vulnerabilities are inherent in the von Neumann architecture or, more generally, in the concept of a programmable device: any legitimate program can be caused to malfunction with the help of another program (malware). The vulnerability seems to be inescapable due to two factors: (a) defense programs designed to neutralize a particular malware turn into a new vulnerability once their algorithms become known to the opponent, and (b) defenses of increasing sophistication quickly reach a level of complexity where their use in real time becomes impossible or impractical. As a result, cyber warfare faces an ever expanding frontier where advantages gained by "outcomputing" the opponent are likely to

Y. M. Yufik (✉)
Institute of Medical Cybernetics, Inc., Potomac, MD 20854, USA
e-mail: imc.yufik@att.net

R. E. Pino (ed.), *Network Science and Cybersecurity*,
Advances in Information Security 55, DOI: 10.1007/978-1-4614-7597-2_5,
© Springer Science+Business Media New York 2014

only be temporary. Second, one can speculate that if the last revolution was brought about by discovering a way to download programs into the device, the next one will commence when a way is found to "download" understanding. Stated differently, emulating mechanisms in the brain endowing the humans with the ability to understand their environment can be expected to yield radical changes in the information processing technology in general and in the methods of cyber warfare in particular. Mechanisms of understanding are the focus of the discussion. With the next revolution pending, the intent is to direct inquiry towards technology that can serve a dual purpose: applying computational power to help analysts to understand complex situations as they unfold, and applying human understanding to help the device to overcome computational bottlenecks.

Discussion is broken into five sections. Section 1 states the problem, Sect. 2 reviews defense tools, Sect. 3 discusses psychology of understanding, Sect. 4 hypothesizes the underlying neuronal mechanisms, Sect. 5 makes tentative design suggestions.

1 The Cyber Warfare Problem

Cyber warfare poses problems of uniquely high complexity, due to an interplay of multiple factors not encountered in conventional warfare [2, 3]. To name a few:

- The World Wide Web is a topologically complex object of astronomical size
- The web has multi-layered architecture
- Conventional warfare uses weapons of limited variety and known characteristics. By contrast, cyber warfare can use a practically unlimited variety of weapons the characteristics and impact of which cannot be known in advance.
- Conventional weapons (except biological and chemical) have restricted radius of impact while cyber weapons can self-replicate and self-propagate across the system.
- Losing a particular function or capability can have a domino effect bringing down other functions and capabilities in the network, etc.

Cyber threats cannot be eliminated: detecting a new type of malware and constructing countermeasures gives temporary protection that remains effective only until the opponent figures out how they work. As a result, cyber defenses can have no fixed "perimeter": every "attack-response" cycle brings on another cycle, thus causing the perimeter to expand. Because of that, cyber warfare has the potential of becoming a "black hole" in the computing universe, consuming ever growing resources without appreciable gains in security.

Figure 1 captures the essence of the problem: Orchestrating the deployment of countermeasures comprised of tools, computational resources and cognitive resources of human analysts, to maximize protection of assets from the current and anticipated threats and to maintain performance throughout the network at or above the levels required by the mission.

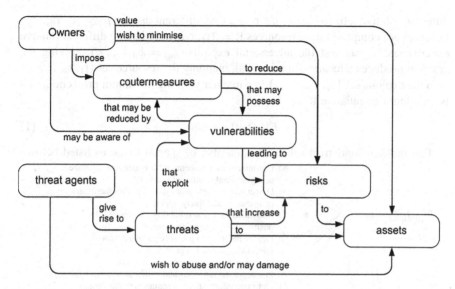

Fig. 1 Countermeasures are deployed dynamically to mitigate the impact of the current and anticipated threats (ISO/IEC 15408-1. Information technology—Security techniques. Part 1: introduction and general model, 2005)

Deployment decisions face a high a degree of uncertainty regarding the situation in the battle space [4]:

- attacker-induced "smokescreens" can generate a flood of false alerts while some of the attack components can be indistinguishable from benign activities,
- attack components can be launched from different locations and separated by large time intervals,
- multiple attacks can be carried out in parallel, etc.

Accordingly, the problem can be expressed as probabilistic resource optimization, as follows:

A set X of N assets $X = \{x_1, x_2,...,x_N\}$ having different relative values A_i ($i = 1, 2,...,N$) is placed on the vertices of network G. Assets are engaged in mission performance which is predicated on maintaining the assets' operational capacity as well as maintaining accessibility between the assets: a path should exist between any two assets in X (more generally, capacity values can be associated with the edges, with the requirement to maintain capacity above some threshold). Assets of the highest value are critical: losing a critical asset aborts the mission. Otherwise, the impact of asset losses is cumulative: from partially degrading to failing the mission, when the sum of losses exceeds some catastrophic level. Assets are interdependent, to a varying degree: degradation (loss) of asset x_i entails degradation (loss) of asset x_j with probability $\omega_{ij} = \omega_{ij}(x_i, x_j)$. Loss or degradation of an asset compromises accessibility (e.g., removes edges incident to the corresponding vertex or degrades their capacities). Attacks are composed of M exploits $Z = \{z_1, z_2,...,z_M\}$ propagating along the edges in G and having different relative effectiveness μ_{ij} respective different assets $\mu_{ij} = \mu_{ij}\,(x_i,\ z_j)$. Defenses are comprised of K resources $R = \{r_1, r_2,...,r_K\}$ having different relative effectiveness λ_{qj} against the adversarial exploits, $\lambda_{qj} = \lambda_{qj}\,(r_q,\ z_j)$. Solving the problem produces allocation matrix $||Q_{ij}||$ mapping the resource set $\{r_1, r_2, ..., r_K\}$ onto the exploits set $\{z_1, z_2,...,z_M\}$ in a manner yielding maximum protection, that is, minimum cumulative losses F

$$F(||Q_{ij}||) \rightarrow min \tag{1}$$

This problem statement subsumes a number of special cases, as listed below:

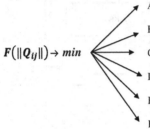

A) Optimize sensor placement for maximum coverage using minimal number of sensors.
B) Detect attacks early at the onset, predict their final composition and likely targets.
C) Maximize detection reliability (minimize false positive – false negative detection rates).
D) Deploy defenses to protect critical assets while minimizing service degradation.
E) Optimize defensive measures vis-à-vis lost/degraded functions and anticipated repair time
F) Optimize information presentation to the human analyst to facilitate situational awareness.
G) Other.

The objective function is comprised of nested multiplicative terms, with the depth of nesting determined by the number of interdependencies taken into account. For example, if simultaneous exploits are presumed to be interdependent (degrading one can cause degradation (reduced efficiency) in some of the other ones), the objective function takes the form

$$F(||Q_{ij}||) = \sum\nolimits_{i=1}^{M} V_i\left\{1 - \prod\nolimits_{j=1}^{M}\left[1 - \beta_{ji}\prod\nolimits_{q=1}^{K}\varepsilon_{qj}(Q_{ij})\right]\right\} \rightarrow min \tag{2}$$

here V_i is the level of threat posed by exploit z_i, β_{ji} is the probability that degrading exploit z_j will degrade z_i, $\varepsilon_{\kappa j} = 1 - \omega_{\kappa j}$, $\omega_{\kappa j}$ is the probability that resource r_q will degrade exploit z_i.

Nested multiplicative terms entail explosive combinatorial complexity. The complexity is vastly increased if cyber networks are coupled with physical networks [5] and if allocation of resources is optimized over the duration of the mission

$$\int_0^T F(||Q_{ij}(t)||)dt \rightarrow min \tag{3}$$

Equation 3 connotes that resource r_i can be allocated to exploit z_j ($q_{ij} = 1$) at some moment in time and re-allocated at some other moment ($q_{ij} = 0$). Dynamic resource allocation enables maneuvering, e.g., accepting a degree of performance degradation early in the mission (e.g., withholding some capabilities to avoid their exposure) in order to prevent unacceptable risks later in the course of mission duration T. Such capability is clearly desirable but, under realistic conditions, computationally infeasible.

In general, procedures for allocating interdependent resources do not have polynomial running times [6]. Approximations are obtained by partitioning the problem into minimally interdependent subproblems: the solution becomes feasible if the degree of correlation between variables across subproblems is sufficiently low. Otherwise, the solution amounts to generating and comparing all the alternatives which is feasible only for toy size problems. Accordingly, if the solution is to be squeezed into small time windows, breaking the problem in Eq. 3 into minimally interdependent subproblems of manageable size is an inescapable necessity. The trade-off is between a narrow window and weak optimization and a wider window and stronger optimization.

To summarize, the complexity of the cyber warfare problem is due to uncertainty and interdependencies inherent on the cyber battle field. Methods for assessing probabilities are laborious but known. By contrast, computational techniques that would allow solving Eq. 3 in realistic, real-time scenarios are not known. The remainder of the paper argues that headway can be made if a cyber-analyst is given the means to interact not only with specific software tools but with an umbrella process orchestrating their deployment. The analyst acts as a "conductor of the orchestra," or supervisor [7–11]. The discussion is prefaced by a brief review of the tools.

2 Tools of Cyber Warfare

Tools can be grouped into three categories: "A—determining what is going on," "B—figuring out what to do about it" and "C—helping the analyst to make sense of A and B." Category A has been receiving the most attention, partly because success in B is predicated on succeeding in A and partly because of the difficulty inherent in diagnosing activities designed to resist diagnosis.

Accordingly, diagnosis employs several approaches. A rich repertoire of techniques ("attack graphs") is centered on analyzing penetration routes through which multi-step attacks can access network devices and exploit vulnerabilities [12, 13]. Attack graph characteristics (e.g., the shortest path) and their derivatives are used as security metrics and probabilistic security estimates [14, 15].

Performance of network devices depends probabilistically on the confluence of conditions in other devices across the network. These interdependencies have been modeled using probabilistic (Bayesian) inference, Petri—net techniques [16] and a number of methods from the arsenal of probabilistic risk analysis [17].

The advantages of diagnosing threats lie in predicting their course. Time Series Analysis, Hidden Markov Model and a variety of statistical prediction methods have been tried out in predicting attack trends [18–20]. Less formal and, it appears, more powerful approaches involve construction of languages optimized for describing attack scenarios. Such descriptions need to be specific enough to allow matching against observable changes in the network and general enough to allow variations in the manner those changes can be carried out. Compromise is reached by defining entities and relations in the network in a probabilistic fashion. For example, the Cyber Security Modeling Language (CySeMoL) supports describing "attack steps" and probabilities of degrading particular assets in each step (the possible to accomplish attribute). Target values and "accomplish" probabilities define the risks and the losses (see Fig. 1) [21]. A variety of "attack" languages focus on particular stages: "detection languages" (e.g., SNORT (Pearson Educational), STATL [22]) specialize in describing early attack manifestations (intrusions).

Notably, there are no comprehensive languages to date in the "figuring out what to do category", that is, no "response language" exits providing a level of support for response scripting and on-line adjustment comparable to that provided by "attack languages" [22].

Finally, tools in the "C—helping the analyst" category are even more scarce than those in the B category. It is recognized that in a cyber battle, as in a conventional battle, success demands decision expediency, that is, ability to complete OODA (Observe, Orient, Decide, Act) cycles faster than the opponent [23]. Decision aids intended to facilitate the analyst through the cycle focus on helping to visualize conditions in the network (Noel et al., [24]; [25] and helping to aggregate (fuse) information). While networks are inherently amenable to visualization, aggregation presents a steep challenge. It has been argued that fusion for cyber warfare is not only in "infancy" but remains largely undefined: "indeed, fusion to provide information and knowledge beyond identifying objects is, perhaps the only certain definition for high level information fusion" The fusion task can be described in technical terms (detecting malicious activity → tracking and correlating alerts → projecting threat and assessing impact) and psychological terms (perceiving → comprehending → anticipating) [26, p. 108]. The next section detours into psychology and visits the subject of comprehension. To motivate the detour, I shall point at the benefits. The anticipated benefit is novel decision technology allowing collaborative solution of Eq. 3 by the analyst (or team of analysts) and decision aid, with speed and accuracy not accessible to either party individually. Solving Eq. 3 in real time enables coordinated erecting, re-shaping and dismantling firewalls limiting access to parts of the network, combined with counter-attacks coordinated across the network (tracing adversarial exploits to their sources and disabling the sources).

3 Situation Understanding

What lies "beyond identifying objects"? Understanding (terms "comprehension" and "understanding" will be used interchangeably) mediates between perceiving an object and anticipating its behavior. Webster's Dictionary defines understanding, or grasp as: "the power to apprehend general relations of particulars." More colloquially, understanding can be defined as "seeing forest behind the tress." How does that work? The question has two aspects: one aspect concerns psychological manifestations of understanding while the other one concerns the underlying neuronal mechanisms. This section addresses the former aspect by referencing findings in the literature and the latter one—by outlining a proposal.

In the context of decision aiding, understanding is often viewed as a component of "situation awareness" that includes "the perception of the elements in the environment within a volume of time and space, the comprehension of their meaning and the projection of their status in the near future" ([27, p. 36]). This definition subordinates comprehension to awareness and connotes the idea that comprehending the meaning of the elements is almost coincidental with perceiving them. The idea is in keeping with Artificial Intelligence (AI) where "understanding" is taken to signify nothing else but possession of knowledge and procedures [28, p. 447] so that, for example, understanding a scene equates to applying procedures, such as neural networks, appropriate for recognizing the elements in the scene. A substantive distinction between AI and the "situation awareness" (SA) construct is due to applicability constraints that are acknowledged in the latter but not in the former: the elements need to be proximal ("within a volume of time and space") and projections reach only into the "near future." The distinction is particularly important in the cyber domain: elements of cyberattacks can be far apart in time and space and their consequences and those of the countermeasures must be projected into the distant as well as the near future. The proximal and the distant can be separated by combinatorial explosions (think of playing chess where projecting just three moves generates about nine million continuations [29]). AI techniques do not scale [30] and thus can be of limited help in either containing such explosions computationally or assisting analysts in the task. On the other hand, the "situation awareness" construct can retain its utility, providing the emphasis is placed on the comprehension component. That is, aiding the analyst in understanding situations unfolding across large networks and over significant time intervals (Eq. 3) necessitates requirements and performance metrics different from those proven adequate under the "proximity" constraints [31]. What are the differences?

Measuring SA involves determining subject's ability to recall "the elements," such as parameters and their values in a simulated process controlled by the subject. During measurement, control is interrupted and subjects are questioned about the elements while facing a blanked display [31]. High SA scores indicate that the elements were noticed and registered in the subject's memory. Invoking the chess example helps in illustrate where SA measures need to be augmented. When players are exposed briefly to a position, master level players recall the pieces and their places on the board faster and more accurately than novices. If the exposure is very brief, even masters might experience difficulties in discerning and/or recalling individual pieces but would still capture the dominant relationship, or the overall "character of the position" (e.g., whites have a weak left flank, etc.) [29, 32, 33]. Moreover, the ability to capture a specific relationship (e.g., checking) varies with the distance and direction on the board (diagonal checks are much less likely to be detected than the vertical and horizontal ones) [29]. These findings are consistent with the Universal Law of Generalization stating that the probability of confusing two items is a negative exponential function of the distance between them in some psychological space [34]. It appears that dimensions of that internal space include temporal and geometrical characteristics of the external elements and, importantly, task-specific relationships between the elements. Performing external tasks (reinforcing a position in chess; hardening a network against anticipated attacks, etc.) is predicated on figuring out routes through the internal space which involves two collaborating mechanisms: noticing and registering the elements populating the space, on the one hand, and capturing the relationships between them, or "seeing forest behind the trees," on the other. Somehow, the collaboration makes possible moving through the space without hitting combinatorial "land mines." SA measures address the first mechanism; additional measures are needed to address the second one. The remainder of this section suggests three measurable constituents of that mechanism and their role in the performance of even the simplest tasks, followed by a discussion of how these constituents can be carried out in the neuronal substrate.

Deriving from the theory and experiments by Piaget [35–39], the suggested constituents are *unification, coordination* and *contraction,* as defined in the following examples (due to Piaget) (A) A child considers his nanny two different persons depending on whether she is in the city where the child lives or in the city where the grandparents live. (B) A child is presented with two glasses having different shapes but equal volumes. As the child watches, the short glass is filled with liquid (say, coke), emptied into the tall glass and filled again. When asked to choose, the child chooses the tall glass "because it has more coke." The experiment can be repeated many times, with the same outcome. (C) Two domino chips are stood on end next to each other. As the child watches, one unit is pushed over and knocks down the other one. Next, multiple units are placed in a row, some at small intervals and some at intervals larger than the length of the unit. When asked to predict what will happen to the last unit in a row if the first one is pushed over, the child proceeds to consider the units consecutively along the row and ultimately fails in either predicting the end result or suggesting re-arrangements that could alter it. Again, repeated demonstrations do not change the outcome.

In all three examples, elements do not combine into a unified memory structure, instead remaining isolated. As a result, the subjects can apprehend neither the fact that some elements remain invariant in a series of episodes (e.g., the traveling nanny), nor the relationships transcending the series. Take the "two glasses" experiment. The child confuses taller with larger, which is a minor mistake. The more fundamental one is a failure to integrate consecutive observations into a unified whole: the (short glass–coke) association must be superposed onto the (tall glass–coke) association so that a unified structure (short glass–coke–tall glass) is produced necessary for the "invariant contents—varying container" relationship to be apprehended.

Coordination is predicated on *unification* and involves interaction between the parts of a whole in such a manner that some integrative property of the whole is preserved when some of the part properties are changing (the notion is consistent with that entertained in [40], I am grateful to Srinivasan of the HRL for pointing that out). An associative whole might resist re-shaping its boundaries which will manifest in restricting one's attention to elements within the boundaries that are relevant to the task at hand while blocking off the irrelevant ones. Another manifestation is the ability to foresee changes in one part of the whole in response to changes in some other part, however remote, without considering changes in the connecting structure. For example, once the domino task is understood, one can foretell the falling of the last unit after the first one is pushed over without mentally tracing, step-by-step as a child does, the fate of the intermediate units. Return again to chess to appreciate the magnitude of advantages yielded by coordination. Before the legal moves are firmly remembered, novices are liable to considering the illegal ones. At the next stage in learning the game, near-term consequences of one's moves are routinely overlooked, such as immediate exposure of own pieces resulting from taking the opponent's piece. However, from that stage on, illegal moves stop slipping in into the thought process. Former world champion Kasparov reported a 15 move look-ahead in one of his games [41]. Without counting the possibilities in a 15 ply look-ahead, one can assume that their number exceeds the number of molecules in our galaxy, which leads to an inescapable conclusion that weak moves don't come to the attention of a master player, no more than illegal ones enter the attention of even a mediocre player [32]. That is, coordination not only carves out small pockets in the astronomically-sized combinatorial volume, but makes possible skipping over swaths of combinations inside the pockets.

Finally, *contraction* consists in reversibly replacing a multitude of elements with a single one, which equates to collapsing a part of the combinatorial volume into a single point (mathematically inclined readers will notice that the definition of contraction echoes the definition of a set (a set is manifold treated as a unity)). Example in Fig. 2 illustrates the work of contraction, in conjunction with the other two constituents of understanding.

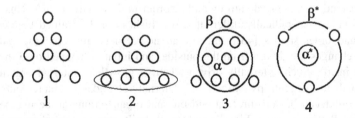

Fig. 2 Understanding the task of turning an arrangement of coins upside down in a minimum number of moves. *1* No unified structure is formed, one can see no other solutions but moving all the coins (10 moves). *2* Partial *unification* reveals a part that can remain intact (six moves). *3* A globally unified structure emerges allowing *coordination* to split the entire arrangement into an invariant and varying parts in a relationship (β revolves around α), *contraction* removes the elements in α from consideration (replaces α with α^*) limiting solution to the elements in β (three moves). *4 Coordination* allows considering the moving all the elements in β^* jointly, as a unit, thus reducing solution to one move

4 Mental Models

Understanding the problem in Fig. 4a involves constructing a model, as in Fig. 4b. Such construction is a form of productive thinking [42] which does not reduce to applying "knowledge and procedures" (what knowledge, what procedures?) or to "pattern recognition" (the circular arrangement α is not immanent to the heap arrangement in Fig. 2a, there is no circular shape hiding inside the heap and waiting to be recognized). Neither is such construction a form of information fusion—the latter term connotes irreversible amalgamation of informational elements, while mental models are amenable to reversible assembly and disassembly, and operations on them are also reversible (e.g., one can alternate freely between α in Fig. 2c and α^* in Fig. 2d without changing either). The notion is important and can be best explained by comparing properties of mental models to those of neural networks.

In neural networks, inputs received in the input layer propagate to the output layer via connections of adjustable strength (weights). Weights get adjusted (trained) for different inputs until some desirable output characteristics are obtained in which case the configuration of weights is said to constitute a model of the input allowing the network to recognize subsequent inputs generalized over the training set. Neural network modeling is irreversible: changing the weights after training vacates the results and disables recognition. The implication would be that one can form and run such models in one's mind but can't think about them without risking distorting or erasing memories of the very events responsible for forming the models. The implication seems to hold for animals where models are formed by conditioning and triggered by environmental cues but not for the humans who form their models in self-initiated and self-directed internal activities (thinking, seeking understanding, as in Fig. 2). A proposal as to how that might work will be presented momentarily, after some brief comments.

Transition in one's memory from fragmented to regularized structures entails reduced entropy and increased order in the underlying neuronal substrate (accompanied by increased entropy and disorder in the environment). This can be carried out in two radically different ways. Fishes can be trained to recognize circles and other shapes [43], and so can neural nets. In both cases adaptive internal changes are driven from the outside, and input is mapped onto the organization inside. Such externally driven adaptive growth of internal organization corresponds to "negentropy extraction." Alternatively, adaptive increments in the internal order can be self-directed and internal organization can be mapped outward onto the input, as in Fig. 2b and 2c: A built-in gestalt mechanism groups coins into horizontal and inclined rows in the former and some other mechanism is responsible for the circular grouping in the latter. This process corresponds to "negentropy production", achieving understanding via constructing and manipulating mental models is presumed to be such a process [44–46].

The key points in the proposal are as follows:

1. A neuronal pool is comprised of a sensory module interacting with the environment, and a control module interacting with the sensory module. Associative links form and grow in strength when sensory neurons repetitively co-respond to contiguous stimuli. Overlapping stimulation in consecutive episodes produces associative chains gradually merging into a connected network.

   Mind the difference between a neuronal pool and a neural network: NN has a fixed organization comprised of fixed sets of neurons and links and responds to different inputs by changing the strength of the links thus changing the value of the output. By contrast, a neuronal pool has a flexible organization and responds to different inputs by selecting different responding neurons thus changing the composition of the output. To underscore: neurons responding to the environment are always the same in NN and vary in the pool. Conditions in the environment change and so do the response compositions.

2. A pool is a self-organizing dissipative system. Due to dissipation, a pool can't respond momentarily but needs time to mobilize, that is, to compose the response. Neurons need time to recuperate. Recuperation period and mobilization rate determine pool's dynamics.

3. A pool has a limited life span. Survival for the duration of the life span is contingent on maintaining energy and nutrient inflows above some critical minima. Survival under changing conditions requires anticipatory mobilization: appropriate neuronal groups (responses are likely to be rewarded) need to be composed and lined up ahead of time. Intelligence is rooted in and is an expression of anticipatory mobilization in the neuronal pool.

4. Anticipatory mobilization involves different mechanisms, conditioning being the foundational, and understanding being the advanced one. Conditioning operates on proximal stimuli, while understanding stretches the capability to allow indefinite separation between the stimuli. Understanding mechanisms are built on top of the processes optimized for handling contiguous stimuli. The range of mechanisms maximizes the chances of survival for the duration of the life span by optimizing composition of neuronal groups, their relative stability, and the order in which they are lined up.

5. Maladaptive neuronal compositions are punished twice: energy and nutrients are withheld, and adjustment work siphons off extra energy. Thermodynamics drives the pool towards seeking stable and successful compositions, optimized dynamically by the internal energy cost–external energy reward trade-offs. The process underlies the subjective experience of apprehending progressively more general and persisting characteristics of the environment (the invariants).

6. Progression from conditioning-based to human level intelligence (understanding) is due to a confluence of two biophysical processes in the neuronal substrate. First, formation of a connected network permeating the sensory module gives it gel-like properties (malleable but firm enough to allow adjustments in some groups to propagate and trigger adjustments throughout the volume).

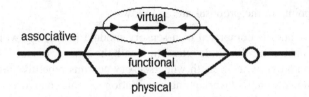

Fig. 3 An associative network is a superposition of three networks: a physical network is formed from neurons having synaptic connections; a functional network forms on top of the physical one as a result of exercising synaptic connections; and a virtual network forms on top of the functional one and comprises neuronal packets. Strengthening of synaptic links has no impact on the physical network, grouping neurons has no impact on the strength of synaptic links

Second, formation of neuronal groups is a form of phase transition in the gel: neurons get packed into cohesive groups (called packets), separated by energy barriers (surface tension) from their surroundings. Surface tension resists the pulling of neurons from their packets of residence. Reduction of surface tension favors merger, while splitting pressure between surfaces (lateral inhibition) pushes packets apart. The interplay produces near-optimal packets connected into a network (called virtual network). Figure 3 illustrates the notion.

7. Grouping is spontaneous. The deliberate (attentive, conscious, self-directed) control interferes to fine-tune the results which includes forming groups of packets and shuffling neurons between packets. Operations on packets are reversible because they have (almost) no impact on the strength of the underlying associative connections. Figure 4 explains the concept.

   Packets shield from combinatorial explosions. For example, if one glass is screened off, the idea of drinking from the other one and looking for more behind the screen doesn't come to the mind of an adult but does occur to children up to a certain age. The same mechanism underlies the ability to steer clear of weak moves in chess, as opposed to considering and rejecting them.

8. The spontaneous and deliberate processes in the pool constitute a self-catalytic loop where increased negentropy production inside the pool entails increased energy intake from the outside providing internal energy surplus sufficient for keeping up or increasing the production. Figure 5 identifies *contraction* as the innermost negentropy-producing operation.

The metaphor of "thermodynamic hand" moving in psychological space underscores the reversibility of cognitive operations: one can think about B after thinking about A and return back to A, or think about A repetitively. The Universal Law of Generalization holds that psychological space is Euclidian; the "hand" pushes elements apart and brings them together (see Fig. 4b) in search for groupings appropriate for the task.

More to the point, one can experience the grouping gestalt in Fig. 2b, organize the solution accordingly, and overcome the gestalt in favor of a different grouping. For example, air traffic controllers were found to be influenced by a grouping gestalt which caused them to detect more readily the impeding collisions between

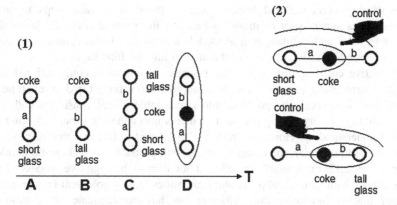

Fig. 4 *1* Two episodes at time A and B might or might not (depending on the duration of the A–B interval) link into a chain at time C following by unification and formation of a packet at time D. *2* Successful unification allows attentive re-grouping (elements are "nudged" to one or the other group) experienced as alternating between two views while maintaining awareness of both. Chaining and operations on packets have no impact on the weights (a and b)

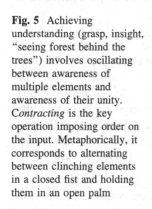

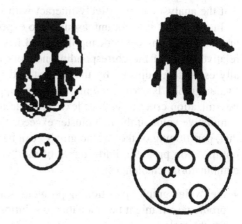

Fig. 5 Achieving understanding (grasp, insight, "seeing forest behind the trees") involves oscillating between awareness of multiple elements and awareness of their unity. *Contracting* is the key operation imposing order on the input. Metaphorically, it corresponds to alternating between clinching elements in a closed fist and holding them in an open palm

aircraft belonging to the same group than between those in different groups. Adequate performance required overcoming the gestalt-based grouping in and forming groups appropriate for the task [47].

5 Supervisory Control in the Cyber Battle Space

To prosecute efficient cyber defense, the analyst needs to understand the situation as it unfolds. Understanding is contingent on forming adequate mental models which, under realistic scenarios, can cause extreme or even insurmountable cognitive burden. At the same time, automating defense puts computer under an overwhelming computational burden. The proposal is to take steps towards establishing a supervisory relationship between the parties: computer helps the analyst to form and validate mental models and analyst helps computer to deal with NP-complete tasks. In the end the analyst has the final say.

Productive collaboration is possible because, in this case, computational and cognitive strategies are in sync: a realistic approach to solving Eq. 3 in real time is to find a way to break the problem into manageable pieces consistent with the human strategy of organizing problem elements into cohesive groups in order to achieve understanding. However, there are significant differences that need to be reconciled. In particular, the computer deals with probabilities and digital entities in a manner that is mathematically efficient but alien to the cognitive process while analyst deals with values and psychological entities (e.g., topological features) in a manner that is psychologically efficient but has no adequate computational expression.

The situation calls for an intermediary. That is, dialogue via a representation format that is comprehensible to the analyst and, at the same time, allows the computer to use human input where it matters most, which is decomposition of the problem. A potential approach could be to extract a subset of the optimization such that the analyst can intuitively interact with it.

In particular, it is not unreasonable to expect that analysts will relate easily to a graph that retains network topology and has vertices and edges weighted by the relative value of the corresponding network elements. Weights are computationally estimated, adjusted by the analyst and vary depending on the mission. The "value graph" is partitioned into non-overlapping clusters under a suitably defined minimization criteria, such as lexicographic minimization over the cluster set (the cumulative value of the first cluster exceeds that of the second one, etc.) Following that, partitioning of the "value graph" can be used to inform decomposition of the optimization problem. Partitioning of the "value graph", however, presents steep combinatorial challenges:

- the "value graph" partitioning problem is a fraction of the original optimization problem so it might be amenable to solution in real time with accuracy sufficient for the task,
- it has been shown that methods of problem partitioning exist that obtain both streamlined decision making and accelerated computation [9, 48, 49,50]. Accordingly, it can be expected that (a) near-optimal partitioning of "value graphs" can be obtained with the help of feedback from the analyst that is intuitive to the analyst and informative to the computation procedure, (b) the results of graph partitioning can be used to obtain weakly optimal solutions of the allocation problem in real time.

Figure 6 illustrates the rough idea.

The format of information presentation in a v-display can change reflecting different levels of information aggregation, with the top level summarizing the main situational trend, such as

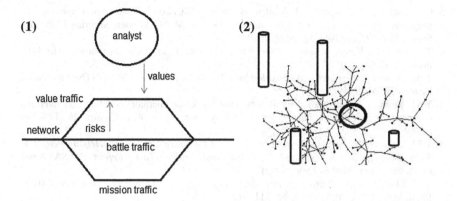

Fig. 6 *1* The network is a superposition of three layers carrying mission traffic, battle traffic (malware and countermeasures) and "value traffic." *2* The "value traffic" display (v-display) shows gains and losses resulting from the flow of allocation decisions in the battle traffic layer. Upon request, the display shows gains/losses resulting from particular decisions, gain/loss propagation across the network, vulnerability spots etc. The layers are transparent, so an analyst can reach down, select a parameter to change and simulate the consequences in the value layer

"There has been growing threat over the last X hours directed at assets A, B, C."

This summary maps directly onto memory organization underlying human situational understanding: it unifies a mass of data acquired over time into a simple, task- relevant structure establishing a relationship between the variable and invariant components of the situation (as in Fig. 2). Reaching such global understanding makes lower- level data elements and events also understandable (e.g., [51]). Analyst's inquiries about such lower-level elements constitute feedback used to streamline and accelerate the underlying computation.

Acknowledgments This research was partially supported by Air Force contract FA8750-12-C-0190 to the Institute of Medical Cybernetics, Inc. The views and opinions expressed in this article are those of the author and do not necessarily reflect the official policy or position of any agency of the US government.

References

1. J. Von Neumann, *The First Draft Report on the EDVAC* (United States Army Ordnance Department and the University of Pennsylvania, PA, 1945)
2. R. Trost, *Practical Intrusion Analysis. Prevention and Detection for the Twenty-First Century.* (Addison Wesley, 2010)
3. J. Andress, S. Winterfeld, *Cyber Warfare: Techniques Tactics and Tools for Security Practitioners* (Syngress, Amsterdam, 2011)
4. S. Cheung, U. Lindqvist, M.W. Fong, Modeling Multistep Cyber Attacks for Scenario Recognition, in *Proceedings Third DARPA Information Survivability Conference and Exposition*, vol. 1 Washington, DC, 2003, pp. 284–292

5. C. Tranchita, N. Hadjsaid, M. Visiteu, B. Rosel, R. Caire, *ICT and power systems: An integrated approach*. eds. by S. Lukszo, G. Deconinck, M.P.C. Wejnen, Securing Electricity Supply in the Cyber Age. Springer, 2010.
6. T. Ibaraki, N. Katoh, *Resource Allocation Problems: Algorithmic Approaches* (The MIT Press, Cambridge, 1988)
7. T.B. Sheridan, *Telerobotics, Automation, and Human Supervisory Control* (The MIT Press, Cambridge, MA, 1992)
8. Y.M. Yufik, T.B. Sheridan, Virtual nets: framework for operator modeling and interface optimization in complex supervisory control systems. Annu. Rev. Control **20**, 179–195 (1996)
9. Y.M. Yufik, T.B. Sheridan, *Towards a Theory of Cognitive Complexity: Helping People to Steer Complex Systems Through Uncertain and Critical Tasks Report* (NASA Ames Research Center, Moffett Field, 1995)
10. Y.M. Yufik, *Memory, complexity, and control in biological and artificial systems* (Proc. IEEE Intell. Syst., NIST, MD, 1996), pp. 311–318
11. S.A. Reveliotis, *Real-Time Management of Resource Allocation Systems: A Discrete Event Systems Approach* (Springer, New York, 2004)
12. S. Jajodia, S. Noel, B. O'Berry, Topological Analysis of Network Attack Vulnerability, in *Managing Cyber Threats: Issues, Approaches and Challenges*, ed. by V. Kumar, J. Srivastava, A. Lazarevic (Kluwer Academic Publisher, New York, 2005)
13. C. Phillips, L.P. Swiler, A graph-based system for network vulnerability analysis, in *Proceedings of the Workshop on New Security Paradigms*, 1998, pp. 71–79
14. N. Idika, B. Bhargava, Extending attack graph-based security metrics and aggregating their application. IEEE Trans. Dependable Secure Comput. **9**(1), 75–84 (2012)
15. L. Wang, A. Singhal, S. Jajodia, Measuring overall security of network configurations using attack graphs. Data Appl. Secur. XXI **4602**, 98–112 (2007)
16. S. Pudar, G. Manimaran, C. Liu, PENET: A practical method and tool for integrated modeling of security attacks and countermeasures. Comput. Secur. **28**, 754–771 (2009)
17. T. Sommestad, M. Ekstedt, P. Johnson, A probabilistic relational model for security risk analysis. Comput. Secur. **29**(6), 659–679 (2010)
18. Q. Wu, Z. Shao, Network Anomaly Detection Using Time Series Analysis. Autonomic and Autonomous Systems, in *Proceedings of the International Conference Networking and Services*, (ICAS-ICNS, 2005)
19. Y. Yasami, S.P. Mozaffari, S. Khorsandi, Stochastic Learning Automata-Based Time Series Analysis for Network Anomaly Detection, in *Proceedings of the International Conference Telecommunications*, St. Peter, 2008, pp. 1–6
20. D.H. Kim, T. Lee, S.O. Jung, H.J. Lee, H. Peter, Cyber Threat Trend Analysis Model Using HMM. (2007), http://embedded.korea.ac.kr/esel/paper/international/2007/1200703.pdf
21. G. Dondossola, L. Pietre-Cambacedes, J. McDonald, M. Mathias Ekstedt, A. Torkilseng, *Modeling of Cyber Attacks for Assessing Smart Grid Security* (International Council on Large Electric Systems, Buenos Aires, Argentina, 2011)
22. S.T. Eckmann, G. Vign, R.A. Kemmerer, STATL: An attack language for state-based intrusion detection. http://www.cs.ucsb.edu/~vigna/publications/2000_eckmann_vigna_kemmerer_statl.pdf. 2000
23. E. Cayirci, R. Ghergherehchi, Modeling Cyber Attacks and Their Effects on Decisions Process, in *Proceedings 2011 Winter Simulation Conference*, 2011, ed. by S. Jain, R.R. Creasey, J. Himmelspach, K.P. White, M. Fu, pp. 2632–2641
24. S. Noel, M. Jacobs, P. Kalapa, S. Jajodia, Multiple Coordinated Views for Network Attack Graphs, in *Proceedings IEEE Workshop on Visualization for Computer Security*, 2005, pp. 99–106
25. K. Lakkaraju, W. Yurcik, A. Lee, NVisionIP: NetFlow Visualizations of System State for Security Situational Awareness. in *Proceedings CCS Workshop on Visualization and Data Mining for Computer Security*, Fairfax, VA, 2004

26. S.J. Yang, A. Stotz, J. Holsopple, M. Sudit, M. Kuhl, High level information fusion for tracking and projection of multistage cyber attacks. Inf. Fusion **10**, 107–121 (2009)
27. M.R. Endsley, Toward a theory of situation awareness in dynamic systems. Hum. Factors **37**(1), 32–64 (1955)
28. H.A. Simon, *Models of Thought*, vol 1–2 (Yale University Press, New Haven, 1979), p. 2
29. N. Charness, Expertise in Chess: The Balance Between Knowledge and Search, in *Toward a General Theory of Expertise: Prospects and Limits*, ed. by K.A. Ericsson, J. Smith (Cambridge University Press, 1991), pp. 39–63
30. J.S. Judd , Neural Network Design and the Complexity of Learning, The MIT Press, 1998.
31. M.R. Endsley, Measurement of situation awareness in dynamic systems. Hum. Factors **37**(1), 65–84 (1955)
32. A.D. De Groot, *Though and Choice in Chess* (Mouton, The Hague, Netherlands, 1965)
33. R. Penrose, A. Shimony, N. Cartwright, S. Hawking, *The Large, the Small and the Human Mind*, (Cambridge University Press, NY, 2007)
34. R.N. Shepard, Towards a universal law of generalization for psychological science. Sci. **237**, 1317–1323, (1987)
35. J. Piaget, *The Psychology of Intelligence* (Harcourt Brace, New York, 1950)
36. J. Piaget, *The Construction of Reality in the Child* (Basic Books, New York, 1954)
37. J. Piaget, *Success and Understanding* (Harvard University Press, Cambridge, 1978)
38. J. Piaget, *The Development of Thought: Equilibration of Cognitive Structures* (The Viking Press, New York, 1977)
39. J. Piaget, *The Grasp of Consciousness: Action and Concept in the Young Child* (Harvard University Press, Cambridge, 1976)
40. W.R. Ashby, *Design for a Brain* (Wiley & Sons, New York, 1953)
41. G. Kasparov, *How Life imitates Chess* (Bloomsbury USA, NYH, 2007)
42. M. Wertheimer, *Productive Thinking* (Harper, NY, 1959)
43. U.E. Siebeck, L. Litherland, G.M. Wallis, Shape learning and discrimination in reef fish. J. Exp. Biol. **212**, 2113–2119 (2009)
44. Y.M. Yufik, Understanding, Consciousness and Cognitive Thermodynamics, published in: Chaos, Fractals and Solitons, Special Edition *Brain and Criticality*, P. Grigolini and D. Chialvo (eds). 2013 (to be published)
45. Y.M. Yufik, Virtual Associative Networks: A framework for Cognitive Modeling. in *Brain and Values*, ed. by K.H. Pribram (Lawrence Erlbaum Associates, 1998a), pp. 109–177
46. Y.M. Yufik, D. Alkon, *Modeling Adaptation and Information Fusion in the Human Nervous System* (Report Office of Naval Research, Arlington, VA, 1998)
47. S.J. Landry, T.B. Sheridan, Y.M. Yufik, Cognitive grouping in air traffic control. IEEE Trans. Intel. Transp. Syst. 2, 92–101 (2001)
48. S.H. Musick, R.P. Malhotra, *Sensor management for fighter applications* (Sensors Directorate, WP AFB, Dayton, 2006)
49. Y.M. Yufik, Method and System for Understanding Multi Modal Data Streams. Patent pending, 2011
50. Y.M. Yufik, T.B. Sheridan, Swiss army knife and Ockham' razor: modeling and facilitating operator's comprehension in complex dynamic tasks. IEEE Trans. SMC **32**(2), 185–198 (2001)
51. J. Kacprzyk, A. Wilbik, S. Zadrozny, An approach to the linguistic summarization of time series using a fuzzy quantifier driven aggregation. Int. J. Intell. Syst. **25**(5), 411–439 (2010)

Design of Neuromorphic Architectures with Memristors

Dhireesha Kudithipudi, Cory Merkel, Mike Soltiz, Garrett S. Rose and Robinson E. Pino

1 Introduction

The advent of nanoscale memristor devices, which provide high-density multi-level memory, ultra-low static power consumption, and behavioral similarity to biological synapses, represents a major step towards emulating the incredible processing power of biological systems. In particular, memristors provide an avenue for designing neuromorphic implementations of artificial neural networks (ANNs) in hardware. The neuromorphic design paradigm, pioneered by Carver Mead [1], seeks to imitate biological information processing using analog or mixed signal circuits, generally leading to higher computational efficiency, lower power consumption, and massive parallelism.

The material and results presented in this paper have been cleared for public release, unlimited distribution by AFRL, case number 88ABW-2013-0820. Any opinions, findings and conclusions or recommendations expressed in this material are those of the authors and do not necessarily reflect the views of AFRL or its contractors.

D. Kudithipudi (✉) · C. Merkel · M. Soltiz
NanoComputing Research Lab, Department of Computer Engineering,
Rochester Institute of Technology, Rochester, NY 14623, USA
e-mail: dxkeec@rit.edu

C. Merkel
e-mail: cem1103@rit.edu

M. Soltiz
e-mail: mgs4513@rit.edu

G. S. Rose
Information Directorate, Air Force Research Laboratory, Rome, NY 13441, USA
e-mail: Garrett.Rose@rl.af.mil

R. E. Pino
ICF International, Baltimore, MD, USA
e-mail: Robinson.Pino@icfi.com

R. E. Pino (ed.), *Network Science and Cybersecurity*,
Advances in Information Security 55, DOI: 10.1007/978-1-4614-7597-2_6,
© Springer Science+Business Media New York 2014

This chapter provides a concise overview of hybrid CMOS/memristor hardware neural networks (CMHNNs), our group's recent research in this area, and the associated design challenges. Section 2 provides background on thin-film memristor operation, and discusses how specific memristor properties can be leveraged to design analog synapse circuits. Section 3 gives an overview of a neural logic block (NLB) designed using memristor-based synapses as well as memristor or CMOS-based digital activation functions. Section 4 provides a brief discussion on different CMHNN network topologies, and Sect. 5 concludes this chapter.

2 CMOS/Memristor Synapse Circuits

Historically, biological neural networks have served as benchmarks for comparing different software and hardware ANN designs. Indeed, the holy grail of many ANN researchers is to emulate the extraordinary performance, robustness, efficiency, and learning capacity of the human brain [2]. However, the neocortex alone contains approximately 20 billion neurons, each making synaptic connections to $\sim 7,000$ (on average) other neurons, for a total of $\sim 1.4 \times 10^{14}$ synapses [32]. Therefore, emulating even a small fraction of the brain's functionality in a hardware based neural network (HNN) requires careful attention to the design of artificial synapses implemented in hardware. In general, each hardware synapse serves three functions: (i) storing weights between neurons, (ii) providing a physical interconnection between neurons, and (iii) facilitating the modulation/programming of weights between neurons. Achieving all of these functions in a low-power, low-area circuit is critical to achieve the high connectivity required for large HNN implementations.

Designs that rely on expensive weight storage circuits, such as capacitor circuits, suffer from large area overheads leading to poor scalability [3]. In the rest of this section, we discuss the design of hardware synapse circuits for a new generation of HNNs based on the integration of CMOS technology with nanoscale memristive devices (CMHNNs). We show that the physical features of thin-film memristors enable the design of compact, low-power synapses that facilitate the implementation of biologically-plausible learning algorithms.

2.1 Memristor Overview and Models

A memristor is a two-terminal passive circuit element that imposes a non-linear relationship between its terminal voltage $v_m(t)$ and the resulting current $i_m(t)$. A simple definition is given by the state-dependent Ohm's law [4]:

$$i_m(t) = G(x)v_m(t),$$

$$\frac{dx}{dt} = f(x, v_m(t)), \tag{1}$$

where x is a state variable, $G(x)$ is the state-dependent conductance (sometimes called *memductance*) of the device, and f governs how x changes over time. We use this definition because it is immediately apparent that $G(x)$ can serve as a synaptic weight between two neurons and can be modulated by adjusting the memristor's state variable. Note that G and f are also functions of a number of constants related to the memristor's initial state and physical characteristics, as well as environmental factors such as temperature [5]. We define a *memristor model* as the tuple (G, f, x). The exact form of G and f, as well as the physical meaning of x all depend on the memristor's physical realization and will also have an impact on how the devices are used to design hardware synapses.

Several thin-film transition metal oxides, such as TiO_2, HfO_2, etc. [6] have exhibited the behavior described in (1). The physical switching mechanism for these devices is often described as a field-assisted uniform drift of defects (e.g. oxygen vacancies). The resulting defect profile directly influences the device's conductance. Assuming the defect drift is linear in the applied electric field, one can derive the widely-used linear memristor model:

$$G(x) = \left(R_{on}x + R_{off}(1-x)\right)^{-1},$$

$$f(x, v_m(t)) = \frac{\frac{\mu}{D^2}v_m(t)}{G(x)}, \tag{2}$$

$$x = \frac{w}{D},$$

where R_{on} is the resistance when $x = 0$, R_{off} is the resistance when $x = 1$, μ is the defect drift mobility, D is the film thickness, and w is the distance that the defects have drifted into the film. This model has had some success in predicting thin-film memristor behavior [7]. However, it does not accurately reflect some phenomena that have been observed, such as non-linear drift, defect/dopant diffusion, temperature variations, and threshold voltages. Some of these behaviors have been captured in better physical models, but they are too complex for large-scale circuit simulation. In order to have high accuracy and fast simulations, we resort to an empirical model based on physical memristor data. The model assumes that the memristance versus applied voltage is a piecewise linear function [8]:

$$M(t_{i+1}) = M(t_i) - \frac{\Delta r \Delta t v_m(t_{i+1})}{t_{\pm(\mp)}v_{m\pm(\mp)}}, \tag{3}$$

where M is the memristance, Δr is the difference between the low resistance (LRS) and high resistance (HRS) states, $t_{\pm(\mp)}$ is the time it takes to switch between the LRS and HRS (HRS and LRS), and $v_{m\pm(\mp)}$ is the negative (positive) threshold voltage.

2.2 Synapse Circuits

Memristors have several important characteristics that make them ideal for use in synapse circuits. Their small footprint—potentially on the order of square nanometers—paired with their integration into high-density crossbars—$1/4F^2$, where F is the crossbar wire half-pitch—enables high connectivity with reduced area overhead. In fact, by stacking layers of memristive switching layers, densities as high as 10^{14} bits/cm^2 can be achieved [9], which approaches the estimated number of synapses in the neocortex. The non-volatility of memristors reduces their static power consumption and thermal stress, which may be important to ensure constant memristor write times across a chip [5]. This is especially important for implementing HNN learning algorithms that require a fixed learning rate. The most important feature of memristors for HNNs is multi-level storage capacity, enabling conductance-coded weight storage in a single device. Finally, the behavioral similarity of thin-film memristors to biological synapses enables simplified implementations of local learning rules [7, 10–12].

In theory, a single memristor is sufficient to provide all three synaptic functions (weight storage, physical interconnection, and weight modulation) [7]. This is especially true for networks of spiking neurons that implement spike time-dependent plasticity (STDP)-based Hebbian/anti-Hebbian learning. Networks of analog spiking neurons with single-memristor synapses are presented in Perez-Carrasco et al. [13], Afifi et al. [10], and a digital implementation is proposed in Ebong and Mazumder [3]. However, these implementations require neurons to output three different voltage levels, complicating the hardware neuron design. Furthermore, the digital implementation requires a complex spiking sequence controlled by a finite state machine. Single memristors have also been used for binary synapse (ON and OFF states only) realization in cellular neural networks [14]. However, additional circuitry is generally needed in a memristor-based synapse design (switches, current mirrors, etc.) depending on which type(s) of learning algorithms (e.g. supervised or unsupervised learning, synchronous or asynchronous learning) and neuron designs (e.g. neuron transfer function, analog/digital implementation) are present in the network. In [15], a series combination of an ambipolar thin-film transistor (TFT) and a memristor is used for synaptic transmission and weight storage in a spiking neural network. The gate of the TFT is controlled by the pre-synaptic neuron, enabling or disabling a constant voltage to pass through the memristor, creating a memductance-modulated current at the input of the post-synaptic neuron. The authors demonstrate learning in a two-neuron network with an average-spike frequency-based learning rule. Kim et al. [16, 17] propose a memristor synapse based on a bridge circuit and a differential amplifier. It can be programmed to implement both positive (excitatory) and negative (inhibitory) weights. It also has good noise performance due to its fully differential architecture. However, it requires three MOSFETs, five memristors, and additional training circuitry (depending on which learning algorithm is being implemented). Another synapse design, presented in Liu et al. [18], incorporates two memristors which can be trained to provide a desired ratio of excitation to inhibition. The synapse also allows bidirectional communication. However, the authors do not include a detailed description of the training circuitry, and the design also consumes a constant static power, which could cause high power consumption in large networks. Another memristor-based synapse design is proposed in Rose et al. [12]. The design operates in subthreshold, resulting in low power consumption. However, the charge sharing technique that the authors employ requires separate pre-charge and evaluate phases of operation, similar to dynamic logic.

Our group has adopted the synapse design from Rajendran et al. [19], which uses a single memristor, an NMOS current mirror, and a local trainer circuit. A generalization of the design is shown in Fig. 1. The memristor is connected to a training circuit or a presynaptic neuron and current mirror depending on the control signal ϕ. When connected between the presynaptic neuron and the current mirror, the memristor modulates the current i_{ipost} as

$$i_{ipost} = Kv_m G. \tag{4}$$

Note that v_m will be a function of the presynaptic output voltage, as well as the current mirror's minimum input voltage. The mirror factor, K, could be used to adjust the resolution of the output current. In our work we have a used a simple CMOS current mirror with $K = 1$. However, cascode mirrors or wide-swing mirrors may be required to achieve necessary output impedances. This will be especially true when several such synapses are connected to the same post-synaptic neuron. The local trainer circuit increases or decreases the memristor's conductance depending on the error signal, which comes from a supervised learning unit. In our simulations, we have implemented the well-known Perceptron learning rule:

$$w_{ij} = w_{ij} + \alpha x_i(y_j^* - y_j). \tag{5}$$

Here, w_{ij} is the weight of the ith synapse connected to the jth post-synaptic neuron. Since our design uses binary output neurons, the weight is simply equal to $Kv_m G$ and the input x_i is '1' when the presynaptic neuron's output is high. Otherwise it is '0'. y_j^* is the expected output of the jth neuron in the Perceptron, and y_j is the actual output. The parameter α is the learning rate and is related to the activity factor of ϕ, the write voltages, and the state of the memristor. It can be shown that this rule will always converge to a global minimum in the weight

Fig. 1 Current mirror-based synapse

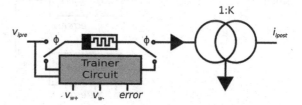

space, provided that there is one. It should be noted that this circuit will consume no static power, aside from the leakage of the trainer circuit, making it scalable to large network implementations.

3 Neural Logic Blocks

The concept of a memristor-based *neural logic block* (NLB) was first introduced in Chabi et al. [20] as a robust alternative to the configurable logic block (CLB) approach to FPGA design. The authors' design is based on a memristor crossbar with threshold logic, bipolar, and binary output neurons. The network is trained to perform linearly-separable functions using the least-mean-squares (LMS) training algorithm, which, from an implementation standpoint, is identical to the Perceptron learning algorithm [21]. Earlier publications also explored neuromorphic approaches to memristor-based configurable logic. A memristor crossbar-based self-programmable logic circuit is presented in Borghetti et al. [22]. The authors demonstrate programming of sum-of-products operations, and self-programming of an AND function. However, only two states (ON and OFF) of each memristor device were used, limiting the scalability of such an approach. Another technique, which is the basis of our NLB design, is proposed in Rajendran et al. [19, 23]. Here, several current mirror-based synapses, like the one shown in Fig. 1, are used as NLB inputs, and a current comparator is used as a threshold logic neuron. However, training is performed offline, so no local trainer is needed. NLBs based on subthreshold synapses are proposed in Rose et al. [12], along with a local/global training circuit that implements a Perceptron learning rule.

In general, an NLB is a self-contained single-layer neural network. If the neurons have a threshold logic binary transfer function, then the NLB is a Perceptron. A block diagram of the NLB considered in our work is shown in Fig. 2. Several synapses (N) identical to the one described in the last section are connected to a common output node, which is the input to a current analog-to-digital converter (ADC). The ADC's output is a $(2^N - 1)$-bit thermometer code. The output of the neuron is a function of the thermometer-encoded input. Since the thermometer code is a one-to-one function of the weighted sum of the inputs $(x_N \ldots x_1)$, then without loss of generality, the output can be described as

Fig. 2 Neural logic block design

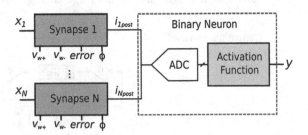

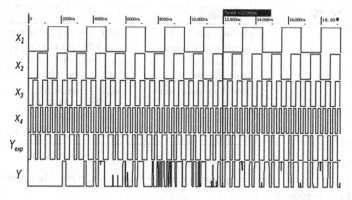

Fig. 3 Training a 4-input NLB to perform an XOR function

$$y = f\left(\sum x_i w_i\right),\tag{6}$$

where f is the activation function. Although not explicitly shown, it should be noted that the ADC can actually be considered part of the activation function.

We have shown in our previous work that training f and the weights in parallel enables the implementation of non-linearly separable functions, such as XOR in a single NLB [24]. Furthermore, the trainable, or adaptive activation function can be implemented using additional memristors, instead of area-consuming flip-flops, which also enables a completely non-volatile network. The major advantage of NLBs in general, and our design specifically, is the large gain in energy efficiency over conventional lookup table-based configurable logic. This is attributed to their low static power consumption and (for our design) ability to implement non-linearly separable functions in a single layer. This last point is demonstrated in Fig. 3. A 4-input NLB is trained to perform a 4-input XOR function. The block is trained using the supervised Perceptron learning rule.

4 CMHNN Topologies

Engineering an artificial neural network is an empirical, ad-hoc process, and is strongly dependent on the ANN's target application. Table 1 lists some common ANN applications and the topologies that have been used for them.

The task of choosing a network topology becomes even more difficult in the case of CMHNNs because of the additional constraints and tradeoffs exhibited by hardware design choices: power vs. speed, size vs. noise tolerance, stability vs. bandwidth, etc. Our group is currently studying two topologies for CMHNNs that demonstrate stark contrast: (i) feedforward Perceptron networks (two-layer and MLP) because of their utility in several applications and the abundance of learning

Table 1 ANN topologies for different applications [31]

Application	Topology
Pattern recognition	MLP, Hopfield, Kohonen, PNN
Associative memory	Hopfield, recurrent MLP, Kohonen
Optimization	Hopfield, ART, CNN
Function approximation	MLP, CMAC, RBF
Modeling and control	MLP, recurrent MLP, CMAC, FLN, FPN
Image Processing	CNN, Hopfield
Classification	MLP, Kohonen, RBF, ART, PNN

algorithms and (ii) random recurrent neural networks (reservoirs) because of their temporal integration ability.

4.1 Perceptron Networks

The Perceptron was introduced by Rosenblatt [25]. His probabilistic feedforward brain model introduced some important concepts still used today, such as an artificial neuron's weighted summation of inputs and non-linear threshold activation function [21]. As we showed in the previous section, a single NLB can be used to implement a Perceptron network. By combining multiple Perceptrons, each trained separately, we can implement multi-layer reconfigurable logic. To demonstrate this, we trained feedforward networks of our NLBs to implement ISCAS-85 benchmark circuits. The networks were trained using the Perceptron learning rule in (5). Our circuits also used the empirical PWL memristor model discussed above, along with 45 nm predictive technology models for the CMOS portions of the design. The energy-delay product results are shown in Fig. 4. RANLB, MTNLB, and ANLB are three different implementations of our adaptive activation function NLB design. The MTNLB (multi-threshold NLB) design [26] outperforms the standard lookup table approach, as well as the threshold logic gate approach (with monotonic activation functions) in all cases. The improvement is attributed to the reduction in NLBs required for a specific function, since each NLB is able to compute both linearly separable and non-linearly separable functions.

4.2 Reservoir Computing

A major limitation of feedforward networks is that they cannot efficiently solve temporal problems, such as prediction, and speech recognition [27]. Random recurrent neural networks or reservoirs, however, naturally integrate temporal signals. The reservoir-based RNN paradigm (reservoir computing) is based on the

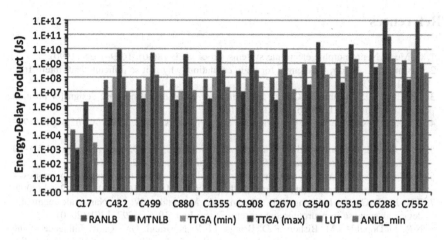

Fig. 4 Energy-delay product results for ISCAS-85 benchmarks implemented using the two proposed designs, RANLB and MTNLB, and a comparison to a TTGA with minimum and maximum memristance values and a standard LUT

work of Jaeger [28] and Maass et al. [29] and can be generalized mathematically by Schrauwen [27]

$$x(t + 1) = f(W_{res} x(t) + W_{inp} u(t) + W_{out} y(t) + W_{bias}),$$
$$y(t + 1) = W_{res} x(t + 1) + W_{inp} u(t) + W_{out} y(t) + W_{bias} \tag{7}$$

where x is the state of the reservoir, u and y are the inputs and outputs of the network, t is time, f is the activation function of the reservoir neurons, and W_i is the weight matrix between i and j ("res" = reservoir, "inp" = input, "out" = output, "bias" = external bias). Training of these networks is simplified by only modifying the output weights. Very simple CMHNN implementations of reservoirs have been demonstrated in Kulkarni and Teuscher [30].

5 Conclusions

In this chapter, we provided a brief overview of CMOS/memristor hybrid implementations of artificial neural networks, with specific attention given to our recent work. This new generation of hardware neural networks represents a major step towards emulation of biological systems. However, several challenges and open questions remain, especially from a circuit design perspective, including: (i) how to provide a good signal-to-noise ratio while maintaining memristor voltages that are below threshold and (ii) how to ensure stability in the presence of noise, especially in the case of recurrent neural networks. More generally, a major challenge is deciding on a synapse design, neuron design, network topology, and other CMHNN parameters given a target application.

References

1. C. Mead, Neuromorphic electronic systems. Proc. IEEE **78**(10), 1629–1636 (1990)
2. R. Cattell, A. Parker, Challenges for brain emulation: why is it so difficult? Nat. Intell. **1**(3), 17–31 (2012)
3. I.E. Ebong, P. Mazumder, CMOS and memristor-based neural network design for position detection. Proc. IEEE **100**(6), 2050–2060 (2012)
4. L. Chua, Resistance switching memories are memristors. Appl. Phys. A **102**(4), 765–783 (2011)
5. C. E. Merkel, D. Kudithipudi. Towards thermal profiling in CMOS/memristor hybrid RRAM architectures, in *International Conference on VLSI Design*, 2012, pp. 167–172
6. R. Waser, R. Dittmann, G. Staikov, K. Szot, Redox-based resistive switching memories— nanoionic mechanisms, prospects, and challenges. Adv. Mater. **21**(25–26), 2632–2663 (2009)
7. S.H. Jo, T. Chang, I. Ebong, B.B. Bhadviya, P. Mazumder, W. Lu, Nanoscale memristor device as synapse in neuromorphic systems. Nano Lett. **10**(4), 1297–1301 (2010)
8. N.R. McDonald, S.M. Bishop, B.D. Briggs, J.E.V. Nostrand, N.C. Cady, Influence of the plasma oxidation power on the switching properties of al/cuxo/cu memristive devices. Solid-State Electron. **78**, 46–50 (2012)
9. K. Cheng, D.B. Strukov. 3D CMOS-memristor hybrid circuits: devices, integration, architecture, and applications, ACM International Symposium on Physical Design, 33–40, 2012
10. A. Afifi, A. Ayatollahi, F. Raissi. Implementation of biologically plausible spiking neural network models on the memristor crossbar-based cmos/nano circuits, in *Circuit Theory and Design, 2009. European Conference on ECCTD 2009*, Aug. 2009, pp. 563–566
11. M. Kulkarni, G. Howard, E. Gale, L. Bull, B. de Lacy Costello, A. Adamatzky. Towards evolving spiking networks with memristive synapses, in *IEEE Symposium on Artificial Life, ser. ALIFE'11*, April 2011, pp. 14–21
12. G. Rose, R. Pino, Q. Wu. A low-power memristive neuromorphic circuit utilizing a global/ local training mechanism, in The 2011 International Joint Conference on Neural Networks (IJCNN), August 2011, pp. 2080–2086
13. J.A. Perez-Carrasco, C. Zamarreno-Ramos, T. Serrano-Gotarredona, B. Linares-Barranco. On neuromorphic spiking architectures for asynchronous STDP memristive systems, in International Conference on Circuits and Systems, 2010, pp. 77–80
14. M. Laiho, E. Lehtonen. Cellular nanoscale network cell with memristors for local implication logic and synapses, in *Proceedings of 2010 IEEE International Symposium on Circuits and Systems*, May 2010, pp. 2051–2054
15. K.D. Cantley, A. Subramaniam, H.J. Stiegler, R.A. Chapman, E.M. Vogel. Hebbian learning in spiking neural networks with nanocrystalline silicon TFTs and memristive synapses, **10**(5), 1066–1073 (2011)
16. H. Kim, M.P. Sah, C. Yang, T. Roska, L.O. Chua, Neural synaptic weighting with a pulse-based memristor circuit. IEEE Trans. Circuit Theory **59**(1), 148–158 (2012)
17. H. Kim, M.P. Sah, C. Yang, T. Roska, L.O. Chua, Memristor bridge synapses. Proc. IEEE **100**(6), 2061–2070 (2012)
18. B. Liu, Y. Chen, B. Wysocki, T. Huang, The circuit realization of a neuromorphic computing system with memristor-based synapse design, in Neural Information Processing, B. Liu, Y. Chen, B. Wysocki, and T. Huang, (Eds) Springer Berlin Heidelberg, **1**, 357–365 (2012)
19. J. Rajendran, H. Manem, R. Karri, G.S. Rose, An energy-efficient memristive threshold logic circuit. IEEE Trans. Comput. **61**(4), 474–487 (2012)
20. D. Chabi, W. Zhao, D. Querlioz, J. Klein. Robust neural logic block (NLB) based on memristor crossbar array, in *Symposium A Quarterly Journal in Modern Foreign Literatures*, IEEE/ACM International Symposium on Nanoscale Architectures, 2011, 137–143
21. A. J. Maren, Craig T. Harston, Robert M. Pap. *Handbook of Neural Computing Applications* (Academic Press, San Diego, 1990)

22. J. Borghetti, Z. Li, J. Straznicky, X. Li, D.A.A. Ohlberg, W. Wu, D.R. Stewart, R.S. Williams. A hybrid nanomemristor/transistor logic circuit capable of self-programming. Proc. Natl. Acad. Sci. USA **106**(6), 1699–703 (2009)
23. J. Rajendran, H. Manem, R. Karri, and G. S. Rose, Memristor based programmable threshold logic array, in IEEE/ACM International Symposium on Nanoscale Architectures, Anaheim, CA, June 2010
24. M. Soltiz, C. Merkel, D. Kudithipudi. RRAM-based adaptive neural logic block for implementing non-linearly separable functions in a single layer, in *Nanoarch*, 2012
25. F. Rosenblatt, The Perceptron: a probabilistic model for information storage and organization in the brain. Psychol. Rev. **65**(6), 386–408 (1958)
26. Mike Soltiz, Dhireesha Kudithipudi, Cory Merkel, Garrett S. Rose, Robinson E. Pino, Memristor-based neural logic blocks for non-linearly separable functions, IEEE Transactions on Computers, 28 March 2013. IEEE computer Society Digital Library. IEEE Computer Society, <http://doi.ieeecomputersociety.org/10.1109/TC.2013.75>
27. B. Schrauwen, D. Verstraeten, J.V. Campenhout. An overview of reservoir computing: theory, applications and implementations, in *Proceedings of the 15th European Symposium on Artificial Neural Networks*, 2007, pp. 471–482
28. H. Jaeger. The echo state approach to analysing and training recurrent neural networks, GMD-German National Research Institute for Computer Science, GMD Report 148, 2001
29. W. Maass, T. Natschläger, H. Markram. Real-time computing without stable states: a new framework for neural computation based on perturbations. Neural Comput. **14**(11), 2531–2560 (2002)
30. M. S. Kulkarni, C. Teuscher. Memristor-based reservoir computing, in *IEEE/ACM International Symposium on Nanoscale Architectures*, ser. NANOARCH'12, 2012, pp. 226–232
31. M. R. G. Meireles, P. E. M. Almeida, M. G. Simões. A comprehensive review for industrial applicability of artificial neural networks. IEEE Transactions on Industrial Electronics, **50**(3), 585–601 (2003)
32. David A. Drachman. Do we have brain to spare? Neurology **64**(12), 2004–2005 (2005)

Nanoelectronics and Hardware Security

Garrett S. Rose, Dhireesha Kudithipudi, Ganesh Khedkar,
Nathan McDonald, Bryant Wysocki and Lok-Kwong Yan

1 Introduction

In recent years, several nanoelectronic device and circuit technologies have
emerged as possible avenues for continued scaling of future integrated circuits
(IC). Nanoelectronics in the context of this chapter is differentiated from CMOS in
that the underlying physics that governs device behavior is in some way fundamentally different from that of classical MOSFET transistors. For example, thin
film metal oxide memristors depend on properties such as electromigration and
other physical state changes that may be considered parasitic for CMOS. While a
variety of nanoelectronic device types exist, each with their own interesting

The material and results presented in this paper have been cleared for public release, unlimited
distribution by AFRL, case number 88ABW-2013-0830. Any opinions, findings and conclusions
or recommendations expressed in this material are those of the authors and do not necessarily
reflect the views of AFRL or its contractors.

G. S. Rose (✉) · N. McDonald · B. Wysocki · L.-K. Yan
Information Directorate, Air Force Research Laboratory, Rome, NY 13441, USA
e-mail: garrett.rose@rl.af.mil

N. McDonald
e-mail: nathan.mcdonald@rl.af.mil

B. Wysocki
e-mail: bryant.wysocki@rl.af.mil

L.-K. Yan
e-mail: lok.kwong.yan@rl.af.mil

D. Kudithipudi · G. Khedkar
NanoComputing Research Lab, Department of Computer Engineering,
Rochester Institute of Technology, Rochester, NY 14623, USA
e-mail: dxkeec@rit.edu

G. Khedkar
e-mail: gck6455@rit.edu

R. E. Pino (ed.), *Network Science and Cybersecurity*,
Advances in Information Security 55, DOI: 10.1007/978-1-4614-7597-2_7,
© Springer Science+Business Media New York 2014

behaviors, there are some characteristics that do appear in common for most technologies. For example, process parameter variability is a common feature for many nanoscale devices. Emerging systems that make use of nanoelectronic devices will either need to cope with or harness nanoscale issues such as variability.

Another field that has gained increasing interest in recent years is that of hardware security. Specifically, IC designers today must cope with a range of potential security and trust issues such as IC counterfeiting, piracy, Trojan insertion and potential side-channel attacks. Many of these hardware security issues can be mitigated through the inclusion of unique identifiers in the design of the IC itself. Unique identifiers for increased hardware security often rely on the inherent variability of device parameters within an integrated circuit. Thus, from this perspective, the inherent variability of nanoelectronic devices is potentially useful in the development of high density security primitives.

This chapter begins in Sect. 2 with an overview of two specific hardware security concerns: trust within the IC design flow and side-channel attacks. Some examples of countermeasures for these threats using CMOS are also provided. Section 3 provides an overview of one particular nanoelectronic device considered in this work for hardware security: the memristor. Circuit level structures for memristive physical unclonable functions that can be used as unique identifiers are discussed in Sect. 4. As another example of uses of nanoelectronics for security, memristor based mitigation for potential side-channel attacks is discussed in Sect. 5. Finally, some concluding remarks are provided in Sect. 6.

2 An Overview of Hardware Security

Since the mid 1970s, information security has evolved from primarily focusing on the privacy of stored and in-transit data to incorporating trust, anonymity, and remote ground truthing. Over this 40-year time frame, the usage scenario of security technologies has evolved from securing physical premises with mainframe computers to securing lightweight, low-cost, and low-power mobile phones, tablets, and sensors. Concurrently, new security metrics such as resiliency against physical and side channel attacks have emerged.

2.1 Threats to Trust: IC Piracy and Counterfeiting

There is an ever-growing industry in counterfeit and recirculated electronics. Experts estimate that nearly 10 % of global technology products include counterfeit components, totally over $7.5 billion dollars in yearly losses to the U.S. semiconductor industry. Even within the U.S. DoD supply chain, over one million components are suspected of being fraudulent [1]. As the cost and complexity of systems increase, so do the losses associated with parts failure. One of the chief difficulties in curtailing this problem is the lack of secure, unique identifiers to verify the authenticity and trust of electronic products. Researchers have proposed Physical Unclonable Functions (PUFs) as a solution.

2.1.1 Potential Countermeasures

PUFs [2–4] are functions that map intrinsic properties of hardware devices into unique "bits" of information, which can then be used as security primitives. With respect to CMOS, these unique bits are derived from device performance differences arising from normal process variability, e.g., propagation delay in a ring oscillator. Typical security primitives include unique identifiers, secret keys, and seed elements in pseudo-random bit generators. With the advances in nanoscale technology, new physical properties have become available to potentially exploit as PUF sources which may allow for more ubiquitous use in the field.

2.2 The Threat of Side-Channel Attacks

Conventional approaches to cryptanalysis exploit vulnerabilities in cryptographic algorithms. A cryptosystem realized on a physical device exudes the information related to the device's operation through unintentional outputs known as side channels. This side channel information can be in the form of execution time, power consumed, or electromagnetic radiation, as shown in Fig. 1. A Side Channel Attack (SCA) exploits the side channel information to gain information about the cryptographic algorithm and secret keys. Based on the side channel information used, various intrusion techniques, such as power analysis attacks [5], timing attacks [6], and electromagnetic attacks [7, 8], have been proposed. Since the seminal SCA attack reported by P.Wright in 1965, SCA attacks have been successfully used to break the hardware or software implementations of many cryptosystems including block ciphers (such as DES, AES, Camellia, IDEA, etc.), stream ciphers (such as RC4, RC6, A5/1, SOBER-t32, etc.), public key ciphers(such as RSA-type ciphers, ECC, XTR, etc.), to break message authentication code schemes, to break the implementation of cryptosystems and protocols, and even to break networking systems. SCAs are growing in significance because they are non-invasive, easy to mount, and can be implemented using readily available hardware [9, 10]. Kocher et al. introduced and demonstrated SCA power attacks in 1999, which extract and analyze the power consumed for different operations to predict the secret key information [5]. In this chapter, focus will be on SCA power attacks due to the sensitivity and variability of this parameter in nanoelectronic devices.

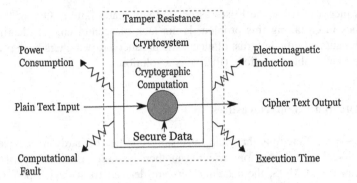

Fig. 1 High-level representation of the side-channel information leakage exuded from a crypto system

There are three fundamental types of SCA power attacks: Simple Power Analysis (SPA) [5, 9, 10], Differential Power Analysis (DPA) [5, 9–11], and Correlation Power Analysis (CPA) [9, 10, 12, 13]. SPA correlates the Hamming weights of the secure data with the power drawn for that computation. SPA requires a detailed knowledge of the implementation of the cryptographic algorithm to observe the variations in the power drawn for different computations [5, 10, 11, 14, 15].

DPA is based on the principle that the bit-toggles from $1 \rightarrow 0$ and $0 \rightarrow 1$ yields significantly distinct power dissipation [5, 7, 13]. DPA needs no detailed implementation knowledge of the cryptographic algorithm and is capable of detecting smaller scale variations that may otherwise be overshadowed by noise and measurement errors while performing SPA [5, 7, 13]. DPA can be countered by introducing random execution patterns or logic/circuit level adjustments [14–19]. CPA is similar to DPA, with additional statistical power leakage models to approximate its power contribution [13]. Current flattening [20] and randomization [21] are adopted to counter the CPA.

In general, a countermeasure to a power attack can be successful if the power drawn is decoupled from the actual computation. Conventional hardware implementations using CMOS have a strong correlation between the power drawn and the associated computation. The countermeasures for these implementations therefore are limited to hiding or masking information, rather than the physical implementation choice itself. Choosing non-conventional and nanoelectronic crypto system implementations, such as Memristors, provides an additional degree of security by their inherent non-linear and reconfigurable nature. The power dissipated in these devices can be decoupled from the actual computation by reconfiguring them to perform different fundamental operations with the same current drawn.

2.2.1 Potential Countermeasures

Countermeasures against side-channel attacks are devised so as to obscure the information leaked through the side-channel. The main goal of the countermeasure is to achieve disconnect between the power profile and the data/operation executed. At the circuit-level, most countermeasures rely on specialized logic styles and devices, for hiding and/or masking. Hiding ensures consistent power consumption activity regardless of the computational state. At the circuit level, dual-rail logic [22], asynchronous logic, current mode logic, and wave dynamic differential logic (WDDL) [19], produce uniform dynamic power for different bit toggles. The major disadvantage of hiding with just CMOS is that it requires capacitive balancing of cells and wires within the chip layout level in order to achieve constant power consumption. These techniques also suffer from excess area and power overheads. An ultra low voltage (ULV) logic using floating gates has been proposed in [16] due to its high speed and low-correlation between the input pattern and supply current, thus making the encryption scheme more resistant to power attacks. Masking at the circuit-level is achieved through dual precharge logic. All of these techniques either incur high area overhead (e.g., 100 % in case of dual rail logic) or require significant design time to maintain capacitive balancing of the cells. Moreover, each technique only covers a specific type of attack.

At the logic level hiding introduces unpredictability in the output power traces and can be achieved using non-deterministic or current flattening techniques. Masking involves randomizing the input data and thereby generating unpredictable side-channel information. Randomization is achieved by shuffling data before computation [23] or inserting non-functional [24] or functional [16] execution steps that intervene with regular computation. Current flattening techniques proposed by Mursen et al. [24] use non-functional instructions intervene with the typical cryptographic algorithm instructions. These additional instructions provide adequate current discharge time to maintain a constant value of the current. For example, an AES based coprocessor [19] has been designed for biometric processing and employs this technique for any bit transaction. However, use of additional instructions leads to a significant increase in the overall power consumption. To address this issue dynamic voltage and frequency scaling have been suggested in [25]. Nevertheless, the power consumption during the non-functional instruction is distinguishable because of its peculiar nature and the original power trace can be recovered by applying a simple time shifting method [24].

Non-deterministic processing overcomes this problem by executing the program in out-of-order as described by May et al. [26]. This diminishes the association between the instructions and corresponding power trace. The cluttered power trace pattern makes it difficult to mount a DPA attack. The out-of-order execution requires judicial run time control to avoid erroneous execution. In [27], a non-deterministic processor with an additional pipeline stage, a mutation unit, is used to randomize the instruction executed while retaining the original execution program. These countermeasures are often not applicable to general-purpose architectures as they bank upon the specialized hardware that requires custom design-flows.

At the system-level, balancing employs another microprocessor to perform a complementary execution in parallel to the normal execution and thereby balancing the overall power consumption [17]. The second processor is used for balancing only when the first processor invokes execution of a cryptographic algorithm or secure application. There are various weaknesses in this technique, which hampers its practicality. First, power dissipation for any two processors is never the same, so power analysis can be refined to mount an attack if the same core is consistently used for cryptographic operation. Second, engaging the second processor only during cryptographic computation reveals the operation time with unique power profile. These weaknesses can be addressed only if the power profile of the device is scrambled enough to hide the operation time information. The inherent variability in memristor-based structures generates a variable power profile even for repeat/redundant instructions, which makes it an attractive fabric to implement. Additionally, the use of memristor devices can address constraints such as manual intervention, high power, and area consumption. For example, the run-time reconfigurability of the memristors can draw current spikes that are not linearly related to the functional operation. This will reduce the need of an application specific hardware to hide or mask side channel information and the associated design-time and fabrication costs.

3 Memristive Devices and Circuits

Memristive devices or resistive RAM (ReRAM) are effectively two terminal electrical potentiometers. That is to say, memristive devices have tunable resistance values yet do not require energy to persist at any resistance state or are non-volatile. By applying the appropriate electrical bias for the required duration, the device may be repeatedly switched between at least two resistance states: a high resistance state (HRS) and a low resistance state (LRS). A SET operation switches the device from the HRS to the LRS; a RESET operation does the reverse. Throughout this chapter, a HRS is a logic '0', and an LRS is a logic '1'.

There is no single memristor device design. Typically, these devices are as simple as metal–insulator-metal (MIM) structures, where the insulating materials have been chalcogenides [28, 29], metal oxides [30, 31], perovskites [32, 33], or organic films [34, 35]. Though the gambit of devices demonstrating the switching behaviors thus described may be understood to be "memristors" [34], the exact switching mechanism, parameters, and style will depend upon the specific material stack. It is worth mentioning that many of the MIM structures studied to date are technically memristive systems [36, 37] due to their nonlinear switching behavior. In this chapter the terms memristor and memristive device are used interchangeably to refer to the family of devices comprising memristors and memristive systems.

Variations in device properties mean that certain flavors of memristive devices may be optimally suited for different applications. Typically, memristive devices considered for digital logic or memory applications are engineered for binary or discrete multi-level states, where abrupt state transitions are desirable. Other devices demonstrate a more analog transition between the two extreme resistance states. In this chapter, we show that both behaviors may be used to develop hardware security primitives.

3.1 Analog Memristors and Write Time

In the simplest analog model, memristors are modeled as two resistors, R_{on} as the LRS value and R_{off} as the HRS value, weighted by a factor α that varies between 0 and 1 over time. In short, the memristance may be written as $M(t) = \alpha(t)R_{on} + (1 - \alpha(t))R_{off}$. While the model is more complex in practice, the idea remains the same.

One method for fabricating memristors consists of placing a TiO_{2-x} layer with oxygen vacancies on a TiO_2 layer without oxygen vacancies and sandwiching them between metallic electrodes [35]. Though conical phase change regions were later shown to be responsible for device switching [38], this device can still be modeled as two series resistors (R_{on} and R_{off}) that represent doped and undoped regions of TiO_2, respectively. In the model, the boundary between the regions (w), the thickness of the active layer, moves between 0 and D as a function of an applied electric field where $\alpha = w/D$. In this way, the transition from the LRS to the HRS is an analog process.

This model has been expanded in [39, 40] to account for variable mobility as described by:

$$M(t) = R_0 \sqrt{\left(1 - \frac{2\eta \Delta R \varphi(t)}{D^2 R_0^2} \cdot \mu R_{on}\right)}, \qquad (1)$$

where constants R_0 is the maximum resistance ($R_0 \approx R_{off}$), Q_0 is the charge required for w to migrate from 0 to D, ΔR is the difference between R_{off} and R_{on}, and η (± 1) is the polarity of the applied voltage signal. The flux $\varphi(t)$ is simply the integral of the applied voltage over the entire usage history of the device:

$$\varphi(t) = \int V_{appl}(t)dt. \qquad (2)$$

Of particular importance to the memristive write-time based PUF considered here is the impact of variations in the device thickness D, similar to the simplified relationship shown in Eq. (1). More specifically, variability in D translates to variations in the read and write-times of the memristor when using the device as a memory cell [41]. For example, a memristor being SET from HRS to LRS will only exhibit a logic '1' output if the SET time (i.e., write time to SET the memristor) is greater than some minimum $t_{wr,min}$. If, however, the SET time is chosen to be at or near the nominal $t_{wr,min}$, then variations in D will dictate that the output is nearly as likely to be a logic '0' as it is a logic '1'. This probabilistic status for the output voltage is undesirable for conventional memory systems but can be leveraged in the implementation of PUF circuits.

3.2 Discrete Memristors

Binary state memristors have only two distinct states, and the transition between the two is typically abrupt. These properties make these devices ideally suited for digital logic and memory elements. Filament creation and rupture is frequently cited as the switching mechanism for these devices.

A PUF circuit was designed specifically for $Al/Cu_xO/Cu$ memristive devices to exploit unique properties detailed in prior research [42]. Unlike most other memristive devices, the $Al/Cu_xO/Cu$ devices switch for any voltage polarity combination, i.e., they are completely nonpolar. The Cu_xO layer is grown via a plasma oxidation process [43]. By virtue of this fabrication process, the oxide thickness and oxygen concentration will vary slightly across the sample. Figure 2 depicts lateral switching (devices in series) of a pair of $Al/Cu_xO/Cu$ devices.

In practice, many memristive materials, including the TiO_2 [44] and $Al/Cu_xO/$ Cu [42] devices considered here, require a forming step to initialize the devices. An elevated voltage is applied across the device to cause the first SET, after which the device can cycle between the HRS and LRS at significantly lower voltages. Prior to this step, the device operates as a regular resistor. The difference in behavior is easy to detect and thus is a prime candidate for tamper detection. In our designs, the memristors are only formed during device provisioning where the PUF challenge response pairs are recorded in a secure environment.

The forming step required to initialize memristor function is of great value for the certification of trust. It serves as a red flag when a device has been activated signaling that the security of the PUF may have been compromised. In addition,

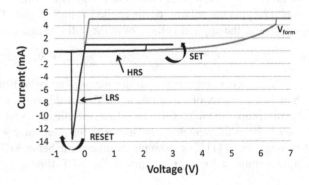

Fig. 2 Representative experimental data from a laterally switched 100×100 μm $Al/Cu_xO/Cu$ memristive devices described in [42]. The *blue curve* shows the required initial forming step. The *black curve* shows typical switching values

such evidence can function as a warning sign alerting the user to suspect fabrication and supply chains.

4 Memristive Physical Unclonable Functions

Some intro to the concept of a memristive PUF...

4.1 Memristive Memory Based PUF Cell

As mentioned for the analog memristors, variations in the thickness D of a memristor leads to variability in the write time (and by extension the read time) of the device. This property is leveraged in the construction of a simple memory-based PUF cell where the SET time t_{wr} is chosen to be the minimum SET time required to switch the memristor from the HRS to the LRS state, $t_{wr,min}$. If the actual SET time of a particular memristor, $t_{wr,actual}$, is greater than $t_{wr,min}$, then the output voltage when reading the memory cell is likely a logic '0'. Likewise, $t_{wr,actual}$ less than $t_{wr,min}$ will likely lead to an output voltage of logic '1'. By choosing the SET time close to $t_{wr,min}$, the likelihood that the output is logic '1' or logic '0' should each be nearly 50 %.

The circuit shown in Fig. 3 is an implementation of a single bit of a memristive memory-based PUF. This circuit is essentially a 1 bit equivalent of the memristive memory presented in prior work [45–47]. Two control signals are used to determine whether the circuit is writing or reading the memristor ($\overline{R}/W$) and, if writing, the polarity of the write (NEG). The circuit works as a PUF by first performing a RESET of the memristor by applying $NEG = 1$ and $\overline{R}/W = 1$ long enough to guarantee the memristor is in the HRS state. Next, a SET pulse is applied for the nominal write time corresponding the $t_{wr,min}$ ($NEG = 0$ and $\overline{R}/W = 1$). After the SET operation, the memristor can be read at the output by applying $\overline{R}/W = 0$.

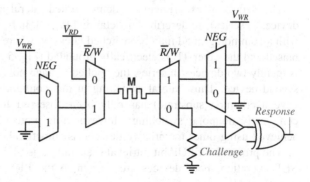

Fig. 3 A 1-bit memristive memory-based PUF cell that leverages variations in memristor write times

The *Challenge* for this particular memristive PUF is applied as an input to an XOR function with the output of the simple memory cell as the other input. The output of this XOR is the *Response* bit of the PUF cell which depends on the *Challenge* and the random output of the memristive memory cell. When the likelihood that the output of the memory cell is logic '1' is 50 %, then the chance that the *Response* can correctly be guessed is equivalent to guessing the outcome of a coin flip.

One way to determine the nominal $t_{wr,min}$ is by running Monte Carlo simulations and producing a histogram of the minimum SET time to SET the TiO_x memristor modeled earlier. Figure 4a, b show plots of the distribution of the SET time for 2 % and 5 % variation in thickness, respectively. From each of these plots, it is clear that the expected minimum SET time for the circuit in Fig. 3 is around 7 μs. Figure 4a, b also show that the standard deviation for the SET time increases with increasing variation in thickness, as is expected. The Monte Carlo simulations were run for 1000 iterations for each variation parameter considered.

Figure 4c, d show the distributions of the output of a read operation to the memory cell after a 7 μs write pulse for 2, 5 % variation in thickness. As was done for the write time distributions, Monte Carlo simulations were run for 1000 iterations using T-Spice. It is clear from Fig. 4 that the likelihood that the output is logic '0' is close to that of logic '1', though it appears a logic '0' is slightly more likely. It can also be seen that as the variation in thickness increases the likelihood for a logic '1' is improved over that of logic '0'.

4.2 Lateral Switching PUF Cell

It has been experimentally demonstrated that $Al/Cu_xO/Cu$ memristive device switching is filament-based [42]. A consequence thereof is a required forming step. The devices will SET at lower voltages only after the forming operation. Thus, by performing a SET operation first, one can test the forming status to verify that all the devices are still in their virgin state. Since the initial switching properties of these devices will be used for the proposed PUF circuit, the ability to verify that the devices have not been previously SET/RESET is critical.

The details of experimentally demonstrated lateral switching (switching two devices in series) are described in detail in [42, 48]. In brief, a pair of MIM devices with a common ground may be switched laterally, where one top electrode (TE) is biased and the other TE is electrically grounded (Fig. 5). While this configuration is merely two devices in series, the applied voltage polarity is reversed across the second device. Thus, lateral switching in this configuration (where two devices have a common substrate) has only been observed for devices demonstrating completely nonpolar switching. However, this protocol in theory may also be achieved using other memristive devices asymmetrically arranged in series.

The protocol for PUF bit retrieval (generation) is SET-RESET. During a lateral SET operation, both devices are written to the LRS; however, after a lateral

(a) **(b)**

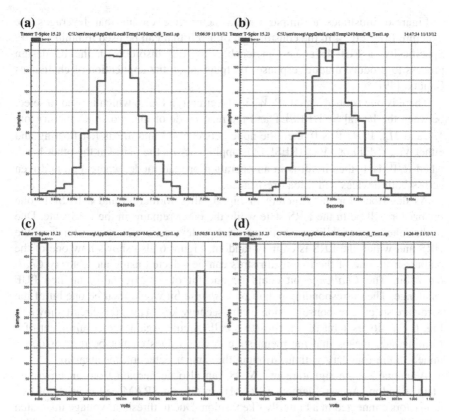

(c) **(d)**

Fig. 4 Monte Carlo simulation results showing the distribution of the write time required to write a logic 1 (**a, b**) and the output voltage (**c, d**) given 2 % (**a, c**) and 5 % (**b, d**) variation in the thickness of the TiO$_X$ memristor. Given such variability in device thickness, the chances that the output of a read operation yields a logic '1' or logic '0' can be made to be close to 50 % by choosing a write time at the center of the write time distribution, in this case 7 μs

Fig. 5 Physical structure of the lateral switching configuration

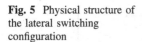

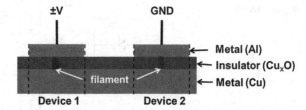

RESET operation, only one of the two devices switches to the HRS. The other remains in the LRS, though, due to process variations, which of the two devices is in the LRS is independent of the device to which the voltage bias is applied. Additionally, over subsequent lateral SET/RESET operations, the persistent LRS device is invariant.

Figure 6 illustrates a simple CMOS-memristive circuit that leverages the structure from Fig. 4 in the construction of a cell that can be used to build a PUF. Specifically, a PUF leverages unclonable physical disorders in the IC design process to produce unique responses (outputs) upon the application of challenges (inputs) [49].

There is one control signal ($R/\overline{W}$) in the circuit in Fig. 6 which is used to select between the lateral SET/RESET and the read mode of the two series memristors *M1* and *M2*. If $R/\overline{W}$ is 0, then the node between *M1* and *M2* is left floating and either V_{WR} (SET) or $-V_{WR}$ (RESET) is applied across the pair. On the other hand, when $R/\overline{W}$ is 1, the circuit is in an operation or read mode, where V_{RD} is driven across both devices and a load resistance.

As described for the structure in Fig. 5, after formation and a RESET, one memristor will be in the HRS state while the other remains in the LRS state. Due to the inherent variability of both memristive devices, which memristor is in the HRS and which the LRS is entirely random. Figure 6 also shows how one of the outputs from one of the two memristors can be selected using an arbitrary *Challenge* bit. The *Challenge* bit could be one bit of an externally supplied PUF challenge. The corresponding output or *Response* bit would then be one bit of the hardware specific response portion of the security key. Thus, the circuit shown in Fig. 6 constitutes 1 bit of a memristive PUF circuit. Again, the requirement of a forming step means that the memristive device values cannot be read or determined in the foundary without tripping the tamper detection mechanism.

A unique device signature in CMOS can also be derived from an unwritten Static Random Access Memory (SRAM) circuit. An SRAM cell consists of two transistors connected in a butterfly like fashion. Due to threshold voltage mismatch caused by process variations, one transistor will be stronger than the other. This mismatch is then used to generate the random signature. However, an attacker in

Fig. 6 A 1-bit filament growth based PUF cell that leverages the stochastic nature of filament formation in some memristors

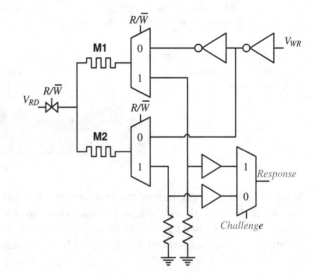

the manufacturing chain can easily read this unique signature and use it to spoof the hardware. Unlike with the memristor-based lateral switching PUF, this tampering is not irrefutable.

5 SCA Mitigation via Memristive Crossbars

In this section, the use of emerging hybrid CMOS/memristor circuits is proposed to mitigate side-channel attacks, particularly power attacks. A highly integrated architecture is proposed which deters the hacker by masking the actual power information associated with the switching activity. The solution is multi-fold as described below.

In this technique, we add different types of noise into the power consumption measurements available to the attacker, reducing the signal to noise ratio (SNR) without fundamentally altering the system architecture. Noise can be generated in the amplitude domain (e.g., by consuming random amounts of power) or in the temporal domain (e.g., by randomizing operation timing). Consider an array of memristors shown in Fig. 7, where each memristor is iso-input in nature. When implementing a specific function each memristor can be used as a bi-level memory element, multi-level memory element, a control knob to monitor power, or as a look-up table based logic element. Also, the underlying physical architecture that interfaces with the CMOS layer is not altered with the reconfigurability (refer to Fig. 7). At any given instance each memristive device can be used for several of these functionalities simultaneously. For example, a memristor can be used actively as a memory element while it is simultaneously being used as a passive monitor. Monitors are used to measure on-chip environmental parameters such as temperature or frequency. When inserted between logic or memory blocks, monitors also add noise or leakage current to the total current drawn. While this specific technique utilizing the reconfigurable feature of memristors does complicate the initial guess for a side-channel attack, a static placement of these devices can potentially be traced to the cracking of the secret key, assuming enough traces are used.

To ensure that the power profile is truly random, monitors are dynamically placed during runtime in unused sections of the crossbar fabric. Since each physical configuration of the monitors and other functional blocks yield a different power signature, this method of dynamic placement creates a unique power profile, which is leaked through the side-channels. Furthermore, dynamic placement of the monitors allows the designer the flexibility to monitor power with higher resolution in areas (encryption and decryption blocks) that are prone to side-channel attacks. This information can be utilized to generate random power profiles even during an ideal state or maintain a uniform profile for a continuous period of time. Active placement of the monitors can be controlled at the operating system level, although hardware solutions for control may also be possible.

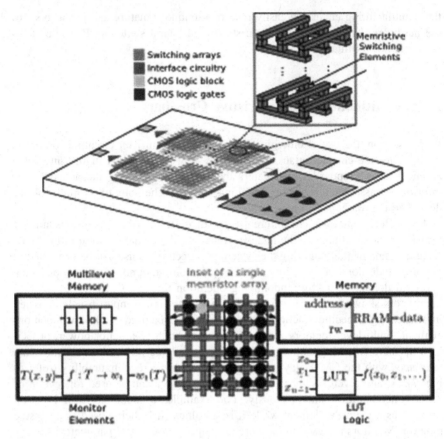

Fig. 7 Block level representation of the memristive switching arrays integrated with CMOS. An inset of a single memristive switching array is shown with heterogeneous functionality

Of all the side-channel power attacks, differential power analysis is of particular concern as it is very effective in identifying the secret key. The hacker uses the information of the power drawn during the switching activity from $0 \rightarrow 1$ or $1 \rightarrow 0$, and then performs statistical analysis of the measured power traces to determine the secret key. To make the set or sequence of operations less dependent on the secret key transitions, active and inactive power drawn for various transitions can be maintained and balanced using multi-level memristive memory elements. Each single memristive memory cell can store 3 data bits, rather than a single bit. The power drawn from this multi-level device will be closer to that of a single transition, whereas the current drawn is utilized for data storage or transmission of multiple bits. This is extremely useful in generating unpredictable behavior.

In general, hardware implementations are most efficient if they can take advantage of some inherent parallelism. With the proposed memristor switching arrays there is an inherent advantage to implement the parallel structures in 3D or

Table 1 Performance metrics comparing pure CMOS to proposed CMOS/memristor structure for a 4-bit full adder

Logic technology	Area (# FETs)	Delay	Standby power	Total power	Active to standby ratio
Super-V_{th} CMOS	110	τ	$15P_{GE}$	$53P_{GE}$	2.53
Sub-V_{th} CMOS	110	$50\,\tau$	$5P_{GE}$	$20P_{GE}$	3
Dual-Rail CMOS	220	τ	$27P_{GE}$	$118P_{GE}$	3.37
CMOS/memristor	20	$\tau/3$	$5P_{GE}$	$16P_{GE}$	2.2

P_{GE} represents the power dissipation of a 2-input CMOS NAND gate

otherwise. The throughput of the combinatorial units can be doubled or increased by a factor of N, by multiplying the number of inner product combinations N times to extract the 128-bit secret key.

Using SPICE-level simulations, Table 1 provides a comparison between the proposed CMOS/memristor technology and existing technologies is made for a simple 4-bit full adder circuit. Though the logic required for the full block cipher in Fig. 6 is more complex than that of a full adder, the results in Table 1 still provide useful insight into the expected gains for using CMOS/memristor technology for SCA countermeasures. Specifically, since the main side-channel to be obscured is power it is important that the ratio between the active power and standby power be as small as possible. As can be seen in Table 1, the CMOS/memristor technology offers the lowest active power to standby power ratio as compared to sub-threshold, super-threshold and dual-rail CMOS designs. From this perspective, it is expected that the CMOS/memristor implementation will offer the strongest level of security against potential side-channel attacks. It is also worth noting that the CMOS/memristor implementation is superior to pure CMOS implementations in terms of area footprint, delay and overall power dissipation.

6 Conclusions

In this chapter, we have highlighted a few novel features of nanoelectronic devices, specifically memristors, and demonstrated how they can be used for constructing security primitives. The features listed in this chapter are based on both experimental and theoretical device research. Using the features listed, device physicists can now engineer nanodevices, not only for memory and logic applications, but also for security applications. Similarly, security researchers can develop mathematical proofs for these security primitives by abstracting the features of nanodevices. Circuit designers can act as a bridge between device engineers and security researchers and construct circuits that will harness these devices to satisfy mathematical strengths. Overall, the idea of using nanoelectronics for security applications will be a new and interesting avenue of research for both electronic and security researchers.

Specifically, memristive devices are good candidates for PUFs due to the heightened effects of process variations on system characteristics. Two specific circuits are discussed which leverage different properties of memristors. First, a memristive memory cell based PUF is presented which leverages variability in the SET time of the memristor. The second memristive PUF considered depends on the ability to read and write two devices laterally, or as a single unit. Preliminary experimental results using Al/Cu$_x$O/Cu devices demonstrate lateral switching wherein, one of two devices becomes fixed in an LRS. Furthermore, preliminary results suggest that which particular device eventually ends up in LRS is random.

The memristor based crossbar is also discussed as a potential countermeasure to side-channel attacks. Specifically, for dynamic power analysis attacks, logic implemented using hybrid CMOS-memristor based crossbars exhibits active power that is very low relative to the standby power of the same system. This low active power allows memristor based systems to generate power profiles that are essentially in the noise and difficult for an attacker to interpret.

More experimental work needs to be done to better understand the switching mechanisms that drive memristors of various flavors, e.g., binary or analog. More experimental data should also be collected for the particular structures considered in this chapter. For example, more measurements of many more lateral switching memristor pairs must be made to better demonstrate the random nature of the lateral switching mechanism. Furthermore, improved device models developed from sound experimentation can be leveraged to better understand the physical parameters of different types of memristors that can be leveraged for PUF operation.

References

1. Inquiry into counterfeit electronic parts in the department of defense supply chain, in *Report 112-167, Committee on Armed Services, 112th Congress, 2nd Session* (United States Senate, U.S. Government Printing Office, Washington, DC, 2012)
2. Y. Alkabani, F. Koushanfar, Active control and digital rights management of integrated circuit IP cores, in *Proceedings of the IEEE International Conference on Compilers, Architectures and Synthesis for Embedded Systems*, 2008, pp. 227–234
3. J. Guajardo, S. Kumar, G.-J. Schrijen, P. Tuyls, Physical unclonable functions and public-key crypto for FPGA IP protection, in *Proceedings of the IEEE International Conference on Field Programmable Logic and Applications*, 2007, pp. 189–195
4. G.E. Suh, C.W. O'Donnell, I. Sachdev, S. Devadas, Design and implementation of the AEGIS single-chip secure processor using physical random functions, in *Proceedings of IEEE/ACM International Conference on Computer Architecture*, (2005), pp. 25–36
5. P. Kocher, J. Jaffe, J. Benjamin, *Differential Power Analysis*, Advances in Cryptology—CRYPTO'99 (Springer, Berlin, 1999)
6. P. Kocher, *Timing Attacks on Implementations of Diffie-Hellman, RSA, DSS, and Other Systems*, Advances in Cryptology—CRYPTO'96 (Springer, Berlin, 1996)

7. D. Agrawal, B. Archambeault, J. Rao, P. Rohatgi, The EM side—channel (s). Cryptogr. Hardw. Embed. Syst. CHES 2002, 29–45 (2002)
8. J.-J. Quisquater, D. Samyde, Electromagnetic analysis (ema): measures and counter-measures for smart cards, in *Smart Card Programming and Security* (2001), pp. 200–210
9. F.-X. Standaert, Introduction to side-channel attacks, in *Secure Integrated Circuits and Systems* (2010), pp. 27–42
10. K. Tiri, Side-channel attack pitfalls, in *ACM/IEEE 44th Design Automation Conference, 2007 (DAC'07)* (IEEE, 2007), pp. 15–20
11. D. Agrawal, R. Josyula, R. Pankaj, Multi-channel attacks. in *Cryptographic Hardware and Embedded Systems-CHES 2003*, pp. 2–16
12. E. Brier, C. Clavier, F. Olivier, Optimal statistical power analysis (2003), http://eprint.iacr.org/2003/152
13. E. Brier, C. Clavier, F. Olivier, Correlation power analysis with a leakage model, in *Cryptographic Hardware and Embedded Systems-CHES 2004* (2004), pp. 135–152
14. C. Clavier, J.-S. Coron, N. Dabbous, Differential power analysis in the presence of hardware countermeasures, in *Cryptographic Hardware and Embedded Systems—CHES 2000* (Springer, Berlin, 2000), pp. 13–48
15. S. Chari, C. Jutla, J. Rao, P. Rohatgi, Towards sound approaches to counteract power-analysis attacks, in *Advances in Cryptology—CRYPTO'99* (Springer Berlin, 1999), pp. 791–791
16. J.A. Ambrose, G.R. Roshan, S. Parameswaran, RIJID: random code injection to mask power analysis based side channel attacks, in *DAC'07. ACM/IEEE 44th Design Automation Conference, 2007* (IEEE, 2007)
17. J.A. Ambrose, S. Parameswaran, A. Ignjatovic, MUTE-AES: a multiprocessor architecture to prevent power analysis based side channel attack of the AES algorithm, in *Proceedings of the 2008 IEEE/ACM International Conference on Computer-Aided Design* (IEEE Press, 2008)
18. S. Guilley, P. Hoogvorst, R. Pacalet, Differential power analysis model and some results, in *Smart Card Research and Advanced Applications Vi* (2004), pp. 127–142
19. K. Tiri, D. Hwang, A. Hodjat, B. Lai, S. Yang, P. Schaumont, I. Verbauwhede, A side-channel leakage free coprocessor IC in 0.18 μm CMOS for embedded AES-based cryptographic and biometric processing, in *Proceedings of the 42nd Design Automation Conference, 2005* (IEEE, 2005), pp. 222–227
20. C. Tokunaga, D. Blaauw, Securing encryption systems with a switched capacitor current equalizer. Solid State Circ. IEEE J. **45**(1), 23–31 (2010)
21. J.-W. Lee, S.-C. Chung, H.-C. Chang, C.-Y. Lee, An efficient countermeasure against correlation power-analysis attacks with randomized montgomery operations for DF-ECC processor, in *Cryptographic Hardware and Embedded Systems–CHES 2012*, pp. 548–564
22. T. Popp, S. Mangard, Masked dual-rail pre-charge logic: DPA-resistance without routing constraints, in *Cryptographic Hardware and Embedded Systems–CHES 2005*, pp. 172–186
23. J. Blömer, J. Guajardo, V. Krummel, *Provably Secure Masking of AES*, Selected Areas in Cryptography (Springer, Berlin, 2005)
24. R. Muresan, C. Gebotys, Current flattening in software and hardware for security applications, in *International Conference on Hardware/Software Codesign and System Synthesis, 2004. CODES + ISSS 2004* (IEEE, 2004)
25. H. Vahedi, R. Muresan, S. Gregori, On-chip current flattening circuit with dynamic voltage scaling, in *Proceedings of 2006 IEEE International Symposium on Circuits and Systems, 2006. ISCAS 2006* (IEEE, 2006)
26. D. May, H.L. Muller, N. Smart, Non-deterministic processors, in *Information Security and Privacy* (Springer, Berlin, 2001)

27. J. Irwin, D. Page, N.P. Smart, Instruction stream mutation for non-deterministic processors, in *Proceedings of the IEEE International Conference on Application-Specific Systems, Architectures and Processors, 2002* (IEEE, 2002)

28. B.D. Briggs, S.M. Bishop, K.D. Leedy, B. Butcher, R.L. Moore, S.W. Novak, N.C. Cady, Influence of copper on the switching properties of hafnium oxide-based resistive memory, in *MRS Proceedings*, vol. 1337, 2011

29. L. Goux, J.G. Lisoni, M. Jurczak, D.J. Wouters, L. Courtade, Ch. Muller, Coexistence of the bipolar and unipolar resistive-switching modes in NiO cells made by thermal oxidation of Ni layers. J. Appl. Phys. **107**(2), 024512–024512-7 (2010)

30. A. Sawa, T. Fujii, M. Kawasaki, Y. Tokura, Interfaces resistance switching at a few nanometer thick perovskite manganite layers. Appl. Phys. Lett. **88**(23), 232112–232112-3 (2006)

31. K. Szot, W. Speier, G. Bihlmayer, R. Waser, Switching the electrical resistance of individual dislocations in single crystalline SrTiO3. Nat. Mat. **5**, 312–320 (2006)

32. J.C. Scott, L.D. Bozano, Nonvolatile memory elements based on organic materials. Adv. Mat. **19**, 1452–1463 (2007)

33. N.B. Zhitenev, A. Sidorenko, D.M. Tennant, R.A. Cirelli, Chemical modification of the electronic conducting states in polymer nanodevices. Nat. Nanotech. **2**, 237–242 (2007)

34. M. Di Ventra, Y.V. Pershin, L.O. Chua, Circuit elements with memory: memristors, memcapacitors, and meminductors. Proc. IEEE **97**, 1717–1724 (2009)

35. D.B. Strukov, G.S. Snider, D.R. Stewart, R.S. Williams, How we found the missing memristor. Nature **453**, 80–83 (2008)

36. L.O. Chua, Memristor-the missing circuit element. IEEE Trans. Circ. Theory **ct-18**(5), 507–519 (1971)

37. L.O. Chua, S.M. Kang, Memrisive devices and systems. Proc. IEEE **64**(2), 209–223 (1976)

38. J.P. Strachan, D.B. Strukov, J. Borghetti, J.J. Yang, G. Medeiros-Ribeiro, R.S. Williams, The switching location of a bipolar memristor: chemical, thermal and structural mapping. Nanotechnology **22**(25), 254015 (2011)

39. Y. Joglekar, S. Wolf, The elusive memristor: properties of basic electrical circuits. Eur. J. Phys. **30**, 661–675 (2009)

40. G.S. Rose, H. Manem, J. Rajendran, R. Karri, R. Pino, Leveraging memristive systems in the constructure of digital logic circuits and architectures. *Proc. IEEE* **100**(6), (2012),pp. 2033–2049

41. J. Rajendran, H. Manem, R. Karri, G.S. Rose, Approach to tolerate process related variations in memristor-based applications, in *International Conference on VLSI Design* (2011), pp. 18–23

42. N.R. McDonald, *Al/Cu$_x$O/Cu Memristive Devices: Fabrication, Characterization, and Modeling*, M.S., College of Nanoscale Science and Engineering University at Albany, SUNY, Albany, NY, 2012, 1517153

43. A.S. Oblea, A. Timilsina, D. Moore, K.A. Campbell, Silver chalcogenide based memristor devices, in *The 2010 International Joint Conference on Neural Networks (IJCNN)*, 18–23 July 2010, pp. 1–3

44. Q.F. Xia, W. Robinett, M.W. Cumbie, N. Banerjee, T.J. Cardinali, J.J. Yang, W. Wu, X.M. Li, W.M. Tong, D.B. Strukov, G.S. Snider, G. Medeiros-Ribeiro, R.S. Williams, Memristor − CMOS hybrid integrated circuits for reconfigurable logic. Nano Lett. **9**, 3640 (2009)

45. H. Manem, G.S. Rose, A read-monitored write circuit for 1T1M memristor memories, in *Proceedings of IEEE International Symposium on Circuits and Systems* (Rio de Janeiro, Brazil, 2011)

46. H. Manem, J. Rajendran, G.S. Rose, Design considerations for multi-level CMOS/nano memristive memory. ACM J. Emerg. Technol. Comput. Syst. **8**(1), 6:1–22 (2012)
47. G.S. Rose, Y. Yao, J.M. Tour, A.C. Cabe, N. Gergel-Hackett, N. Majumdar, J.C. Bean, L.R. Harriott, M.R. Stan, Designing CMOS/molecular memories while considering device parameter variations. ACM J. Emerg. Technol. Comput. Syst. **3**(1), 3:1–24 (2007)
48. J. Rajendran, R. Karri, J.B. Wendt, M. Potkonjak, N. McDonald, G.S. Rose, B. Wysocki, Nanoelectronic solutions for hardware security (2012), http://eprint.iacr.org/2012/575
49. B. Gassend, D. Clarke, M. van Dijk, S. Devadas, Silicon physical random functions, in *Proceedings of the ACM International Conference on Computer and Communications Security* (2002), pp. 148–160

User Classification and Authentication for Mobile Device Based on Gesture Recognition

Kent W. Nixon, Yiran Chen, Zhi-Hong Mao and Kang Li

1 Introduction

Intelligent mobile devices, sometimes called smart mobile devices, have become the boilermaker of the electronics industry in the last decade. The shipment of smartphones and tablets triples every six years, and is anticipated to reach a volume of 1.7B in 2017 [1]. Mobile devices have become widely used in many aspects of everyday life, spanning uses such as communication, web surfing, entertainment, and daily planning. Also, almost all mobile devices are now equipped with some sort of cellular network connection and/or Wi-Fi modules, with the majority of them being connected to the internet regularly or from time to time.

A recent statistics shows that around 66 % of intelligent mobile devices store a moderate to significant amount of private data [2]. The typical sensitive information on the mobile device includes personal contacts, passwords, social networking info, employment and salary info, online banking account, a local copy of cloud-based documents, etc. In [3], a survey shows that once the device is lost or an unauthorized access is wrongly granted, those data will be immediately under severe security risk.

Password and keypad lock are two commonly implemented methods of mobile device security protection. Interestingly, although many consumers agree that the

K. W. Nixon (✉) · Y. Chen · Z.-H. Mao
University of Pittsburgh, Pittsburgh, PA, USA
e-mail: kwn2@pitt.edu

Y. Chen
e-mail: yic52@pitt.edu

Z.-H. Mao
e-mail: zhm4@pitt.edu

K. Li
Rutgers University, Piscataway, NJ, USA
e-mail: kl419@rci.rutgers.edu

R. E. Pino (ed.), *Network Science and Cybersecurity*,
Advances in Information Security 55, DOI: 10.1007/978-1-4614-7597-2_8,
© Springer Science+Business Media New York 2014

data security is important, only 49 % of them have enabled at least one security mechanism on their devices [3]. The currently available security solutions require extra operations on top of normal operations, e.g., typing a password before being granted access to the device, which may be the primary cause for such a low adoption rate. Some biometrics solutions, such as fingerprints, etc. often require additional hardware support, leading to even lower utilization.

In this chapter, we present our latest research on mobile user classification and authorization based on gesture usage recognition. Different from other biometrics solutions, gesture is uniquely defined by the interaction between a user and their smartphone's hardware design, and evolves throughout the whole life of the user. The required sensing technologies, such as touch screens, gyroscopes, accelerometers, gravity sensors, etc., have been equipped in most modern mobile devices. It offers an evolutionary security protection over the operation time of the mobile device.

2 Security Solutions of Mobile Device

2.1 Conventional Security Solutions

Mobile devices are now used in many aspects of our daily life, either if used for productivity purposes or just pure entertainment. Only 22 % smartphones are strictly for personal usage. The rest of smartphones are used either exclusively for business purposes, or for both personal and business matters. Following the increase of data storage capacity, more and more data are temporarily or even permanently retained on the smartphone, which incurs severe concerns about data security.

The security solutions employed in the cell phone have existed for a long time, being utilized much before the smartphone. Although it is one of the most common security solutions, a password is also the most vulnerable one. Due to the restrictions implied by their usage model, smartphone passwords usually comprise of a combination of only 4–6 numeric digits. The number of unique passwords available with so few characters is very limited, and therefore easily hacked by existing software or hardware methods. The use of less robust passwords, e.g., using birthday or other meaningful combination as the password, puts the data integrity of smartphone at immediate risk.

Very recently, another security solution has been introduced into smartphone design, namely, the track pattern. The track pattern requires the user to trace across some nodes shown on the screen using their finger, following some preset sequence. Obviously, as the number of available sequences may be more than the number of combinations available to password users (e.g., 10000 for 4-digit), it is assumed to offer better protection than a traditional password. However, the track pattern scheme may not in fact be as safe as the manufacturers claim. As the

Achilles' heel of the track pattern solution, the operation of tracing the pattern will leave am inerasable track on the screen due to the dirt and oils present on the user's hand. If the track is discovered and interpreted, the track pattern can be easily hacked.

Although many users agree that the data safety is a real concern during the operation of mobile devices, very few people turn on either of the above two protections. We believe that this is because the current password and track pattern protection methods each require an additional operation on top of the normal operations of the smartphones. In some cases, these additional operations can be annoying or even generate some safety issue, e.g., unlocking the cell phone during driving.

2.2 Biometrics Solutions

Note that the password and track pattern only record simple information like digit combination or node sequence. Hence, they are very easy to be replicated. In some applications that require a higher level of security, many biometrics solutions have been developed. These solutions are generally linked to some unique biological characteristics of the user's.

Two poplar biometric security solutions for general data protection are iris and fingerprint scans. The unique patterns of an individual's iris and fingerprint can be recognized by a device as the only passport to normal access and operation. Very recently, human facial features are also used as a method of authentication. Apparently, the information carried by biometric characteristics is irreproducible and the corresponding security solution must be much safer than simple passwords or track patterns. However, these solutions require a special device in order to extract the biometric information, leading to additional hardware cost and the inconvenience of the usage model.

Moreover, these biometric solutions are based on static information per se. In military applications, such solutions may not be able to protect the safety of both the data and the user simultaneously. For example, fingers may be removed from a captured soldier in order for enemy agents to be able to pass through a fingerprint recognition system. In extreme cases such as these, linking the security protection to the biologic characteristics actually increases the threat to a user's safety and still could not necessarily protect the data.

2.3 Requirements of Mobile Device Security Solutions

We have summarized the desired characteristics of a security solution for intelligent mobile devices in the list below:

- The solution must be coded by unique characters.
- The code must be irreproducible.
- The code must be difficult to hack, or the hacking cost must be sufficiently high.
- The solution must be able to protect the safety of both data and users.

As we shall show in the next section, the recognition of gesture usage satisfies all the above conditions while offering an adaptive and evolutionary security solution.

3 Why Use Gesture to Protect Your Smartphone

3.1 Why Gesture is Unique

The human hand is one of the most complex organs. Figure 1 shows the muscle–tendon structure of a human hand [4]. The subtle operation of the gesture is affected by the biological features of the hand, such as finger length, rotational radian, fingertip area, and palm size, as well as the muscle–tendon parameters. These biological features are irreproducible across users, and changes over time with the user's medical condition and age [5].

Gesture denotes the way we use our fingers to control the mobile device. The combination of the above mentioned biological features are responsible for the diversity of gesture. In fact, the finger dynamic behaviors can be more complicated since these behaviors may also have a "feedback" impact on the musculoskeletal

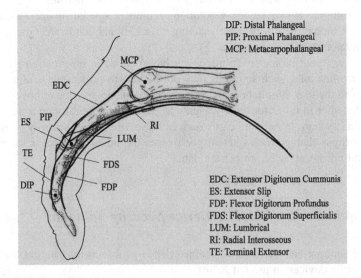

Fig. 1 Muscles of a human finger [4]

Fig. 2 Muscle-tendon parameters such as flexor and passive torques are both user-dependent and behavior-dependent [5]

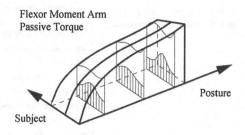

parameters in Fig. 2, which may make gesture a more attractive method for user identification.

Besides the physical and biological factors of the user, e.g., the structure and the biological features of human hands, much human–machine interaction information is also generated during gesture operations. Some examples include:

- Preferable Finger: Even for the same gesture operation, different user may have different preferable finger to execute it.
- Preferable Screen Area: Every user has his/her preferable screen area even for the same operation due to the unique comfortable room between the human hand features (e.g., palm size) and the cell phone's physical configuration (e.g., screen size).
- Wrist/Finger Angle and Holding Position: They are usually linked to the physical and medical condition of the user and reflected as the preferable screen area and other sensing parameters.

The information associated with gesture operations is much richer than just the physical motion of the fingers on the touch screen. There are many sensing parameters describing many details on gesture operations, including:

- Tap Frequency: The frequency of taping the finger on the touchscreen.
- Sliding Speed: The speed of the finger moving on the touchscreen.
- Sliding Angle and Radian: Describing the curve shape of finger motion on the touchscreen.
- Pressure Strength: Describing how hard the finger pressing on the touchscreen.
- Pressing Time: Duration of the finger press on the touchscreen.
- Contact Area: Area of the finger contact with the touchscreen. It usually depends on the fingertip area and is correlated with the pressure strength.

Obviously all of the parameters related to the gesture operations are more than what we listed above. If we can recognize the gesture patterns by using these parameters and link it to a unique user identity, we will be able to classify the user and provide a more reliable security solution.

Table 1 Sensors integrated on samsung galaxy nexus S

Name	Model (if applicable)
3-axis accelerometer	KR3DM
3-axis magnetic field sensor	AK8973
Light sensor	GP2A
Proximity sensor	GP2A
Gyroscope sensor	K3G
Rotation vector sensor	Software sensor
Gravity sensor	Software sensor
Linear acceleration sensor	Software sensor
Orientation sensor	Software sensor
Corrected gyroscope sensor	Software sensor

3.2 Sensing Capability of Mobile Devices

Modern mobile devices embed many sensors that can be used to extract the required parameters. Table 1 lists the sensors integrated on the Samsung Galaxy Nexus S. Of the ten sensors, only the first five are really of interest, as the last five are "software sensors"—virtual sensors created using data collected from the first five dedicated hardware sensors.

Figure 3 shows a typical sensor data collection over time for a 3-axis accelerometer and gyroscope. The sensed data patterns change during the motion of the user, e.g., on the bus or other moving target, walking, lying on the bed, etc. These sensors can also detect the position in which the device is being held. Considerations such as this will cause the same gestures completed on the same mobile device to show some differences under different scenarios.

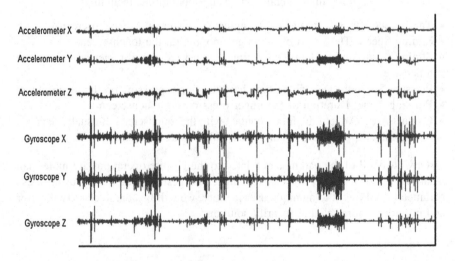

Fig. 3 Accelerometer and gyroscope data collection

Thanks to the latest display technology, touchscreens can record many finger motion parameters during the operation of the mobile device, such as movement track, pressure strength, and contact area, etc. These parameters can be combined with the those detected by still other sensors to derive much additional information.

3.3 Functional Mechanism of Gesture Security Solution

Figure 4 shows the gesture-based security solution for mobile devices. The interaction between the mobile device and user generates the gesture operations whose parameters are recorded by the embedded sensors and touchscreen. Once collected, these data are sent to the pattern recognition algorithm in order to determine whether the operation was executed by a rightful and authorized user of the device, or by an attacker. If the pattern is confirmed, the minute differences between the current operation and the stored patterns will be characterized and used to further tune the stored verification patterns. Otherwise, the system needs to identify whether the pattern is simply a new for an existing user, or is in fact being used by an invader. A local database for all trained patterns will be retained on the mobile devices for fast access. However, if the number of patterns is too large, only the most frequent patterns will be stored on the mobile device while the others may be kept in the cloud. Note that the recognition of gesture can be executed in the background and shall not interrupt the standard function of gestures in other applications. Our target is that the continuous execution of this gesture recognition algorithm will not noticeably decrease the overall performance of the device.

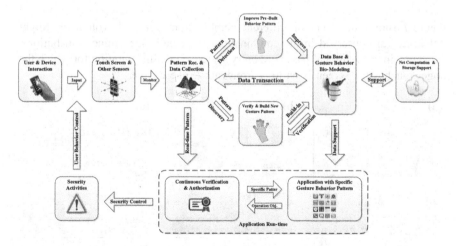

Fig. 4 System overview of gesture-based security solution

4 Experimental Results

4.1 Experimental Setup

To demonstrate the potential of the proposed gesture-based security scheme, we analyze the plethora of additional information that can be gathered during conventional password entry. An application was created to collect the gesture information generated by a user while entering their password into a device. Three different combinations are tested to analyze the impact of the distance between the keys in the password, as shown in Fig. 5. The key distance varies from the minimum to the maximum.

4.2 Pressing Time Distributions

We first compare the pressing time distributions (PTDs) of three volunteers for the combination of 1245, as shown in Fig. 6. The timeline begins when the finger touches the screen for the first digit. The distribution of the first digit is the time when the finger leaves the screen. Starting with the second digit, the two distributions denote the time when the finger touches and leaves the screen, respectively. The distances between two adjacent digits are almost equal while the distributions increase following the rising of the order of the digits. However, the patterns of three volunteers are very different, as we can observe.

4.3 Combination Dependent PTD

Figure 7 shows the PTD of different combinations for each volunteer. Slight differences are observed in the combinations, especially the timing distributions. However, in conventional password entry systems, the time duration for which the finger touches the screen is entirely irrelevant to the device.

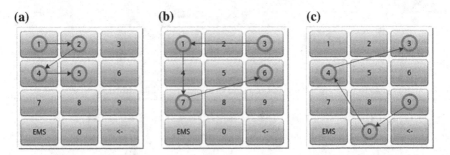

Fig. 5 Three tested combinations. **a** 1245. **b** 3176. **c** 9043

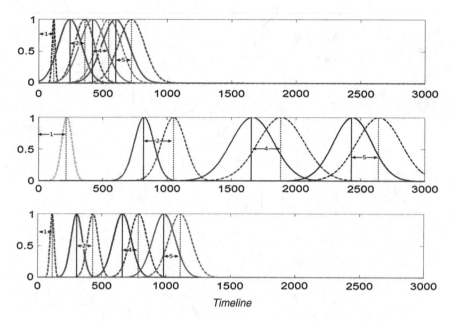

Fig. 6 Pressing time distributions of three volunteers

4.4 Contact Area and Pressure

Table 2 summarizes the statistics of the contact area for all volunteers. There are significant differences of the gesture behavior across the volunteers with different gender and age. Also, the contact area deviations are tiny for the different combinations for one user, indicating stable gesture behavior. However, the differences between the users are large.

Table 3 summarizes the statistics of the pressure strength for all volunteers. Similar conclusions can be drawn for all the users and password combinations. Very interestingly, the physical features or even the gender information of the users may be exposed by the above data. For example, a large contact area indicates a large fingertip, or generally an adult (compared to a child) or a male user (compared to a female user).

5 Future Research

Although our initial results proved the informational richness of the gestures used to interact with mobile devices, many technical obstacles still need to be overcome in order to enable the technology. For example, a proper learning algorithm must be developed to recognize the gesture patterns. The capacity of the pattern

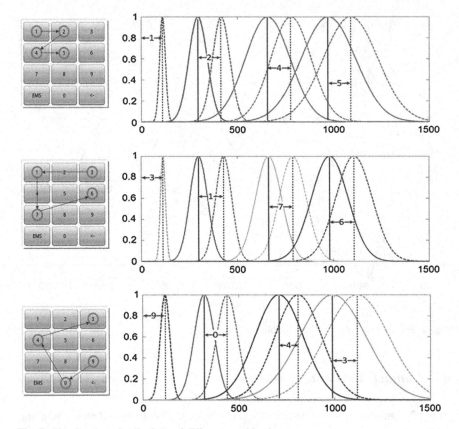

Fig. 7 Pressing time distributions of different combinations

Table 2 Summary of contact area statistics

Average	Comb1	Comb2	Comb3	Std. Dev.
Volunteer 1	0.1707	0.1762	0.1715	0.0029
Volunteer 2	0.1456	0.1541	0.1529	0.0046
Volunteer 3	0.1977	0.2008	0.1998	0.0015
Std. Dev.	0.0212	0.0191	0.0192	

Table 3 Summary of pressure strength statistics

Average	Comb1	Comb2	Comb3	Std. Dev.
Volunteer 1	0.5293	0.5561	0.5411	0.0134
Volunteer 2	0.4611	0.5013	0.4910	0.0208
Volunteer 3	0.5668	0.6100	0.5796	0.0222
Std. Dev.	0.0436	0.0443	0.0363	

database should not be too small or too large, and should equally balance accuracy and processing time considerations. The algorithm must be neither under- nor over-trained. As another example, the relationship between the gesture variance and the ambient parameters (e.g., daytime or nighttime) and user's medical/racial conditions will be also analyzed. Finally, a portable recognition platform must be established to allow transferring the gesture database from one device to another one, each of which may have different hardware configurations, so that the security solution is not limited to a specific device.

6 Conclusions

In this work, we reveal the demands on an adaptive and evolutionary security solution for intelligent mobile devices. As one promising candidate, gesture offers almost all desired characteristics of ideal data protection schemes. Its resilience to hacking is guaranteed by both the complexity of the gesture operation as well as the unique nature of a user's biological features. We then discuss the method to recognize the user's gesture as the data security protection, including the software and hardware requirements for doing so. Our analysis demonstrates that gesture offers a promising and user-friendly solution for mobile device data protection, though many technical obstacles still need to be overcome.

References

1. OVUM Object Analysis, "Ovum expects smartphone shipments to reach 1.7 billion in 2017 and Android to dominate as OS", http://ovum.com/section/home/
2. AVG Mobile Survey, "Smartphone security—a study of U.S. consumers," http://resources.avg.com/
3. Symantec, Smartphone Honey Stick Project, http://www.symantec.com
4. K. Li and X, Zhang, A novel two-stage framework for musculoskeletal dynamic modeling: an application to multi-fingered hand movement. IEEE Trans. Biomed Eng. 56(7), 1949–1957 (2009)
5. K. Li, X. Zhang, A probabilistic finger biodynamic model better depicts the roles of the flexors during unloaded flexion. J. Biomech. 43(13), 2618–2624 (2010)

Hardware-Based Computational Intelligence for Size, Weight, and Power Constrained Environments

Bryant Wysocki, Nathan McDonald, Clare Thiem, Garrett Rose
and Mario Gomez II

1 Introduction

Nanoelectronic research is enabling disruptive enhancements for computational systems through the development of devices which may enable autonomous human-system interactions. Intelligent platforms that collaborate with their human operators have the capacity to enhance and compliment the human capability, adding resiliency and adaptability to current systems while reducing operator tedium. Programmable machines are limited in their ability to address such fuzzy combinatorialy complex scenarios. Neuromorphic processors, which are based on the highly parallelized computing architecture of the mammalian brain, show great promise in providing the environmental perception and comprehension required for true adaptability and autonomy.

It is challenging to anticipate the capabilities and forms that future computationally intelligent systems may take, but recent advancements in nanotechnology, biologically-inspired computing, and neuroscience are providing models and enabling technologies that are accelerating the development of such systems. Thus, as illustrated in Fig. 1, neuromorphic computing design is truly a multidisciplinary technology area. This chapter briefly examines many of the breakthroughs and

B. Wysocki (✉) · N. McDonald · C. Thiem · G. Rose
Information Directorate, Air Force Research Laboratory, Rome, NY 13441, USA
e-mail: Bryant.Wysocki@rl.af.mil

N. McDonald
e-mail: Nathan.McDonald@rl.af.mil

C. Thiem
e-mail: Clare.Thiem@rl.af.mil

G. Rose
e-mail: Garrett.Rose@rl.af.mil

M. Gomez II
Department of Engineering, Science and Mathematics, State University of New York
Institute of Technology, Utica, NY 13502, USA

R. E. Pino (ed.), *Network Science and Cybersecurity*,
Advances in Information Security 55, DOI: 10.1007/978-1-4614-7597-2_9,
© Springer Science+Business Media New York 2014

Fig. 1 Multidisciplinary
nature of neuromorphic
computing design

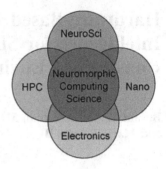

trends that are influencing the development of solid-state neuromorphic systems
and surveys a variety of approaches to hardware design and fabrication.

2 Neuromorphic Processors over Software for Embedded Systems

Advanced sensor platforms provide an overwhelming and exponentially growing
supply of data. Geospatial systems for example, generate crushing amounts of data
that choke communications channels and restrict or prohibit real-time analysis.
This has created a strong demand for autonomous pattern recognition systems [1–
3]. Software applications running on high performance processors using conventional Turing computation have been very successful but are failing to efficiently
scale to the required processing speeds while maintaining a viable power budget
for embedded systems. Increased performance and bandwidth within the Turing
formalism can alleviate some of the strain but are only part of a more sophisticated
solution. Massively parallel neuronal architectures are achieving the reduced size,
weight, and power (SWAP) requirements needed for embedded platform applications and are benefiting from advances in supporting fields. This fundamentally
different approach, often referred to as neuromorphic computing, is thought to be
better able to solve fuzzy perception and classification problems historically difficult for traditional, von Neumann-based computers.

The focus of processor enhancement took a historic shift from increased clock
speeds to multi-core design around 2005. Power and performance issues redirected
the trend of transistor scaling that lasted over 30 years towards multi-threaded parallel architectures [4]. But the utility of multi-core systems is limited by the restricted
scaling of parallelization (Amdahl's argument), software optimization complexities,
and memory storage and retrieval issues [5, 6] such that more and more researchers
are investigating unconventional computing methods for the next era of enabling
technologies. Researchers look to the physical world for inspiration for information
processing algorithms and architectures based on the thermodynamic, chemical,
quantum, and biological processes for non-classical approaches to computation and
sense making.

Neuromorphic computing seeks to mimic brain functions and efficiencies through parallel operation, reconfigurability, high density, and low power consumption. High performance computer based simulations can emulate brain function, but are incredibly inefficient and slow. Additionally, the energy requirements of the fetch, decode, and execute model of von Neumann computers grows exponentially with scaling and limits the use of Teraflop systems to only the largest research facilities. Modern supercomputers typically consume 4–7 MW of power, enough to power over 5,000 homes, with the largest supercomputers consuming over 12 MW of power and approaching $10 M in yearly operational costs. The largest obstacle to extreme scale computing is in fact power and it has been estimated that Exaflop systems would require approximately 1 GW [7]. This is roughly the average output of the Hoover Dam or a nuclear power plant for operation. With the growth of big data and the internet, data centers are also becoming significant draws on U.S. and global power production at approximately 2.0 % and 1.3 % respectfully in 2010, with many single centers having nearly a 100 MW capacity [8]. Even at the other extreme of computer processing, the average desktop uses between 50 and 160 W of power which is still too much for most embedded systems. The need for increased intelligence with reduced power consumption in computational systems is great for all levels of computing.

The DOD is the nation's largest energy user and energy efficiency, as a force multiplier, is a top priority. The demand for autonomy and computationally intelligent systems provides increasing pressure for the development of next generation processors. Many of the platforms which could benefit most from onboard high performance computing are constrained by size weight and power limitations. This is the ideal niche for hardware-based neuromorphic processors which compliment traditional computing with newfound adaptability and resilience on a reduced requirements budget. Enabling technology advancements are spurring neuromorphic hardware development and expanding the engineer's toolbox.

3 Hardware Design Considerations

The availability of novel technologies spurred by advancements in nanotechnology, hybrid fabrication, and neuroscience has provided circuit designers with previously unavailable options. This section examines some of those options.

3.1 CMOS Implementations

The pressure for increased speed and efficiency is pushing the limits of CMOS technology which are expected by many to hit-the-wall by 2022. The diminishing financial return on investment for the continued scaling of CMOS technology may force mainstream fabrication to stall with 350 mm wafers at the 22 nm node. While there are numerous technological possibilities for the extension of CMOS, there has been increased focus on a number of beyond CMOS candidates such as memristors, nanowires, nanotubes, and graphene, in addition to entirely new architectures such as nano-crossbar and 3D networks on a chip [9]. These alternatives, however, must compete with the fact that CMOS based VLSI is a relatively inexpensive and well proven technology that will continue to see drops in fabrication prices on older nodes from 0.5 μm down to 45 nm. CMOS platforms also benefit from the availability of standardized modeling software for system design such as Spice and Verilog-A. CMOS is so engrained into processor design that it proves very difficult to avoid and is present in many neuromorphic schemes. Reconfigurability is a drawback for CMOS that can be overcome through the development of hybrid technologies where memristive devices add some plasticity.

3.2 Hybrid Designs

Many of the new nano-enabled technologies are not strictly compatible within current CMOS fabrication lines. They require front-of-line or back-of-line processing and often have special packaging needs. Great effort is underway, however, to bring these capabilities into mainstream fabrication lines.

Memristors for example, take numerous forms of construction based on a variety of technologies, including resistive random access memory (ReRAM), phase change RAM (PCRAM), magnetoresistive RAM (MRAM), and spin-transfer torque MRAM (SST-RAM). Thus, a wide range of performance characteristics can be achieved using varied material and architectural designs, all demonstrating the characteristic pinched I-V hysteresis and non-volatility. Such variety makes standardization slow, but gives neuromorphic circuit designers expanded fabrication options.

There are two critical tasks for successful memristive device integration with CMOS: manufacturability and usability. Concerning the former, the devices to be used must consist of materials that are permitted inside a CMOS foundry, which further restricts the materials allowed in the front end of line (FEOL) as compared to the back end of line (BEOL). All the processing steps needed to make the devices' structure must be scalable to fabricate devices en masse. Lastly, all of the devices must be functionally identical (though some applications may actually exploit device non-uniformities). Part of the difficulty of manufacturing memristive devices is that the physics of device switching is not well understood at nanometer size scales. In particular, ReRAM (of which PCRAM is a subset) may be composed of binary metal oxides, chalcogenides, or perovskites, among other materials, and switch due to filament formation, vacancy migration, phase change, or other processes [10]. At these scales, small variations in the device size or material composition often have large effects upon subsequent device switching parameters.

However, because of this variety of materials and mechanisms, different device resistance values, switching voltages, and switching times are available to the circuit designer. When considering the appropriate device metrics of reliability and endurance that must be attained, one must first consider the intended use of the device. For von Neumann computing applications, if these devices are to replace Flash or SRAM, then endurance cycles of about 10^6 and write speeds of a couple tens of nanoseconds must be achieved, respectively. Even if memristive devices cannot meet these requirements, SWAP savings may still be achieved by strategically replacing some transistors in a circuit. For devices used in neuromorphic applications, the range of addressable resistance values and the operative voltages will be more critical than the write speed. Because of these varied ends, there will likely be a variety of memristive device "flavors" available to the circuit designer in the future.

3.3 Analog Versus Digital

Analog systems allow the device physics to do the computation which more closely mimics the nonlinear dynamics of biological systems. This is not unsimilar to structures built on Bayesian classification and regression, artificial neural networks, or kernel methods which often utilize non-linearities to project the input into higher-dimensional spaces for improved separability. Some main advantages to analog designs, besides their compatibility with natural signals, are their large bandwidth and real-time performance. But such complex circuits are difficult to reliably design and lack reprogramability limiting their applications [11]. Subthreshold analog circuits are additionally sensitive to device variations, which can result in reduced bit resolutions after digital conversions; fortunately many of the applications for neuromorphic systems do not require such exactness. Indeed, they are designed to solve the fuzzy computational problems too ambiguous for traditional processing. In reality, few computational systems are completely analog and there are nearly always points where the information is most useful in one form or the other. There are several practical benefits of digital systems that are well known. Chief among these benefits is their insensitivity to noise followed by comparatively cheap fabrication costs and high device densities [12]. The most probable near term realizations of large scale neuromorphic circuits will continue to use both analog and digital circuitry for what they respectively do best.

4 Selected Design Examples

There are many hardware implementations of neuromorphic computing. This section takes a high level look at a few selected architectures.

4.1 Crossbar Nanomatrix

Crossbar architectures containing memristive nodes have the potential to serve as high-density memory fabrics, or, with the addition of active components, as crossbar-based nanoelectronic circuits. Such designs, however, are not without their challenges which become more evident at the nanoscale with half-pitches approaching 10–20 nm. Issues with leakage currents, energy consumption, active device alignment, addressing and fan-outs are typical [13, 14]. Never-the-less, crossbars remain a viable and promising method for memristor implementation as these challenges are addressed.

Two different crossbar implementations of CMOS-memristor architectures are considered as examples in this section as detailed in Fig. 2. The first architecture is simply a crossbar array where each crosspoint consists of a single memristor. Since memristors are essentially resistive devices, such a crossbar is typically limited due to sneak path currents and other parasitics. In the case of a typical neuro-morphic computation, however, vector inputs can be represented by voltages applied simultaneously to all input rows of the crossbar. If the columns of the same crossbar each drive an independent load, then the resulting output will be determined by the sum of the currents through each memristor in the column resulting from the product of the respective input voltages and memristance values. If analog input voltages represent elements of an input vector and the memristance

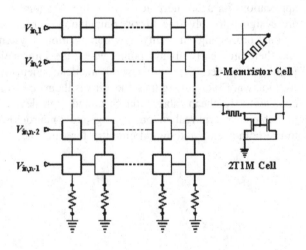

Fig. 2 Nanomatrix illustration. Cells (*white* boxes) can be single memristors or the patented 2T1M cells

values correspond to the weights of associated synapses, then the output is a vector resulting from a neuromorphic computation.

A second, competing implementation is based on a 2T1M cell, two transistors and one memristor, at each crosspoint in the array. The two transistors in this basic cell comprise a current mirror which isolates the input voltages on each row from the output seen on the columns. Such an arrangement is certainly not as dense as the purely memristive crossbar, but it does benefit from improved noise margins and the lack of back-driving (unregulated current). Thus, one implementation provides the greatest benefits in terms of performance (memristor only) while the other is likely more robust (2T1M). These structures (1M, 2T1M) are examined in more detail in the next section.

4.2 CMOS/Memristor Hybrid

Here the single memristor (1M) crossbar and the 2T1M structures are examined as examples of CMOS-memristive neuromorphic implementations. In prior related work, the 2T1M structure has been explored for use as part of a basic threshold logic cell [15–18]. Specifically, the memristive element in a 2T1M cell is used as a weight for a particular input and several of these cells are connected at their outputs such that the resulting currents can be summed together. A P-type metal-oxide-semiconductor (PMOS) current mirror cell is used to drive a reference current against the summed current representing the weighted sum of the inputs. The reference current in this case would function as the threshold for this CMOS-memristive threshold logic gate. The functionality of this circuit as a threshold gate can be described as:

$$Y = \begin{cases} 0 \ if \ \sum \frac{V_i}{M_i} < I_{ref} \\ 1 \ if \ \sum \frac{V_i}{M_i} \geq I_{ref} \end{cases}, \tag{1}$$

where M_i is the memristance value for corresponding input V_i with the "mem-conductance" (memristor conductance) $1/M_i$ representing the weight for that input [15, 18]. The reference current I_{ref} represents a threshold such that if the weighted sum is less than the threshold, then the output is low. Otherwise the output is high. In the 2T1M based implementation, the final comparison between the weighted sum and the reference current is made by the fighting of the two currents on the output column [16]. Figure 3 shows a three input threshold logic gate constructed with 2T1M cells.

Threshold logic gates like that shown in Fig. 3 were explored as part of a collaboration between AFRL and the Polytechnic Institute of NYU (NYU-Poly). In that earlier work, an algorithm was developed to map Boolean logic functions to a circuit comprising of 2T1M based threshold logic gates by adjusting the weights for the desired functionality. For the 3-input circuit shown in Fig. 3, the energy and delay have been calculated and can be seen in Table 1 for several possible

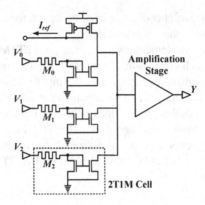

Fig. 3 A 3-input current mirror threshold gate which uses the memristors as weights and I_{ref} as the threshold [15]

Boolean functions. The average delay is around 1 ns with only a few femto-Joules of energy consumed per function [15].

Mapping Boolean logic functions to threshold logic gates as shown in Fig. 3 is useful for early benchmarking of such circuits and comparing the performance to pure CMOS implementations. However, threshold logic actually supersedes Boolean logic and is particularly useful for non-Boolean operations, such as in neuromorphic approaches to computing. Thus, the threshold logic circuit shown in Fig. 3 can be particularly useful for implementing artificial neural networks.

As a dense alternative to CMOS-memristive 2T1M cells, the basic crossbar structure has been considered for several years for implementing nanoelectronic logic and memory [15, 19–21]. Clear advantages for a crossbar array with only 2-terminal memristors at each crosspoint is the density and low expected power consumption. The basic crossbar array, however, does suffer from several disadvantages, namely, sneak path currents, back-driving and a potential intolerance to noise. Back-driving results from the fact that each memristive crosspoint is resistive such that current can flow in either direction and with no built-in way to differentiate the output of the circuit from the input. That said, back-driving and noise intolerance can be mitigated by carefully designing the peripheral circuitry.

Table 1 Characteristics of some 2T1M based gates mapped to boolean logic [15]

Function	Memristance (MΩ)			Energy (10^{-15} J)	Transistor count	Delay (ns)
	A	B	C			
AB	1.2	1.2	–	9.10	12	1.97
A + B	0.6	0.6	–	8.98	12	0.982
ABC	1.2	1.2	1.2	6.8	14	1.92
A + B + C	0.4	0.4	0.4	7.29	14	0.637
AB + BC + CA	0.8	0.8	0.8	6.65	14	1.19
AB + AC	0.6	1.2	1.2	6.93	14	0.928

Several examples of CMOS-memristive memory design exist where the CMOS peripheral circuitry is designed to take full advantage of the dense, memristive, crossbar array memory [20, 22, 23].

The issue of sneak paths is particularly troubling for memristive crossbar array based circuits. Without inserting diodes (1D1M) or transistors (1T1M) along with memristors within each crosspoint of the crossbar, the output of any column is dependent not only on the selected memristive cell but on every other device in the array as well. Essentially, the unselected devices in a crossbar memory, for example, can be considered as being in parallel to the selected device such that current will flow through and depend on both paths. To make matters worse, for larger sized arrays the resistance of the unselected path (consisting of all unselected devices) becomes smaller such that more current will flow through the unselected devices than through the selected device. Put another way, as the crossbar array size grows the outputs depend less on the selected devices than on the unselected devices [22].

While the use of 1T1M or 2T1M cells provides solutions to the sneak path issue it is still worthwhile to explore ways to make use of the high density single memristor (1M) cells. As mentioned earlier, neuromorphic computing is one example that could actually leverage the sneak path issue as a feature. Specifically, consider the case where the rows of a crossbar are driven by voltage inputs representing the input vector and the columns are all pulled through independent loads such that the voltages at the columns represent the output vector. In this same array, the memconductance of each memristive crosspoint would be used to represent a synaptic element. Since the columns are each connected to a load at their outputs, the only sneak paths of concern are those within each column. These particular sneak paths would be used to add the products of input voltages and respective memconductance values to form the output voltage or the respective output of a neuron.

In addition to the circuits required to implement recall functions, consideration must be given to how these circuits are to be trained, or, in some cases, directly configured. In prior work, several supervised training techniques have been explored for circuits such as the 2T1M threshold logic circuit shown in Fig. 3 [24–26]. The primary issue with implementing supervised training techniques in hardware is that a significant amount of area overhead could be required. Memristor burn-in, when required, must also be considered.

One approach for mitigating this hardware overhead is to clearly separate the training circuitry between what must be included with each synapse or weight element (local trainer) and what can be shifted to more of a neuron level (global trainer) [26]. The reasoning here is that there are many more weight elements than there are neurons so reducing the overhead for what is required per weight element can have a dramatic improvement on the overall overhead. In the hardware based cases considered thus far, the training circuitry would follow a supervised training model.

As an example of the training circuitry considered in prior work for 2T1M based threshold logic circuits [15], the stochastic gradient descent model [26] has been employed. Stochastic gradient decent works by incrementally adjusting the

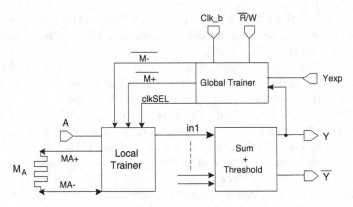

Fig. 4 Fully connected perceptron with training circuitry [26]

weights (i.e., memristance values) based on whether or not the output matches a given expectation. An illustration of the local/global trainer division for this particular approach is shown in Fig. 4. Here, the memristor would be part of a weighting cell (e.g., 2T1M) and the "Sum + Threshold" block would represent the threshold based neuron circuitry.

For the implementation in Fig. 4, training consists of providing a series of expected outputs (Y_{exp}) along with the corresponding input vectors (A, B, C, etc.). The global trainer in this case compares the expected output Y_{exp} with the actual output Y and determines if a weight adjustment is required and, if so, whether the weight should be increased or decreased. Two signals, in addition to a clock, are communicated from the global trainer to all local trainers whether adjustments are required and in what direction: $M+$ is high for a required positive adjustment while $M-$ is high for a required negative adjustment. Each local trainer receives $M+$ and $M-$ signals and if the input (A, B, C, etc.) associated with the particular weight is high then the local trainer will adjust the memristive weight by a set increment depending on $M+$ and $M-$. In Figs. 5 and 6, this is demonstrated with an example for a 3-input 2T1M based threshold logic gate being trained to implement the function $Y = AB + AC$ [26].

4.3 ASIC Artificial Neural Networks

The implementation of real-time ANNs is ideal for pattern recognition in platforms with severe Size, Weight and Power (SWAP) constraints. These restrictions practically rule out traditional software approaches which often run too slowly due to the inherent serial nature of von Neumann architectures or require high performance processing for operation. While nano-enabled neuromorphic architectures are extremely promising, their realization will take time. Meanwhile, there exist commercially available technologies that offer partial solutions. In particular,

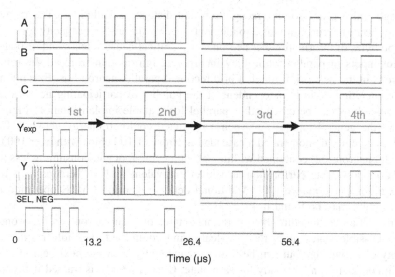

Fig. 5 Simulation results for training a configurable logic block to $AB + AC$ logic functionality [26]

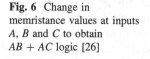

Fig. 6 Change in memristance values at inputs A, B and C to obtain $AB + AC$ logic [26]

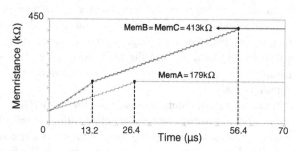

parallel processing capabilities are afforded by FPGAs [27] and general-purpose computing on graphics processing units (GPGPUs). The focus of this section is concerned with the recent availability of application-specific integrated circuits (ASIC) based on zero instruction set computing (ZISC). They offer a reduction in footprint and power with native support for massively parallel operations. It is technically feasible, using current state-of-the-art fabrication techniques at the 22 nm CMOS node, to manufacture ASIC ANN chips approaching 500,000 parallel neurons.

A recently avaiable, fully parallel, silicon-based neural network chip (CM1K) developed by CogniMem Technologies Inc., and based on IBM's earlier series of ZISC chips [28, 29], provides a scalable network for pattern recognition. The CM1K is configured with two available types of non-linear classifiers: a Radial Basis Function Network (RBF) and a K-Nearest Neighbor classifier (KNN). The chip possesses 1024 neurons, each with its own memory for trained signature storage and a processor for recognition and distance calculations. The memory within every neuron contains 256 elements, each with an 8-bit capacity for a total of 256 bytes of information per neuron. The identical neurons learn and respond to vector inputs in parallel while they incorporate information from all the trained neurons in the network through a bi-directional parallel neuron bus. Execution of the recognition logic is independent of the number of participating neurons, and multiple chips can be cascaded in parallel for scalable implementation. Figure 7 shows the general topology of such a restricted coulomb energy network. CogniMem recently demonstrated a cascaded network of 100 chips with over 100,000 parallel neurons, all contained within 1/10 of a cubic foot and consuming less than 20 W of power yet performing at a level equivalent to 13.1 Teraops of pattern recognition performance [30]. Additional details regarding CM1K operation and architecture may be found in [31, 32].

In such an architecture, the operational status of each neuron can be in one of three possible states: idle, ready-to-learn, and committed. The idle neurons are empty of knowledge but can be trained sequentially with the next neuron in the chain configured in the ready-to-learn state. Once a neuron is trained it becomes committed and any pre-existing influence fields are adjusted to accommodate the new knowledge. During recognition, the input vector is passed to all the committed neurons in parallel, where it is compared to the stored vector or trained prototype. If the distance between the input vector and a neuron prototype falls within the influence field, the neuron "fires" generating local output signals consisting of fire flag, signal distance, and category type. In the case that no neurons fire, the input signal can be used to train the ready-to-learn neuron with the unrecognized signature. This provides the means for recognition and training to be accomplished simultaneously.

Fine tuning of neuron sensitivity for a specific signature can be manually adjusted by adjusting the neuron active influence field or the distance from ideal, where a neuron will still recognize the target as a specific category. The distances can be calculated using one of two norms: the Manhattan method,

Fig. 7 Neural network diagram. Each input node accepts a maximum of 256 elements ($\times$N), each with 8-bit resolution. These are fed in parallel to up to 1024 neurons. All recognition events are passed through to the output layer with the associated category and confidence level

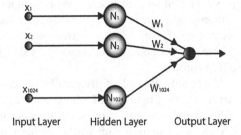

$$D_{Man} = \sum_{i=1}^{n} |V_i - P_i|,$$ (2)

where D_{Man} is the sum of the differences between n dimensional vector signatures V_i and P_i, or the Lsup method,

$$D_{L\,sup} = Max|V_i - P_i|,$$ (3)

where D_{Lsup} is the maximum separation of V_i and P_i. A neuron fires when the input vector lies within a specified distance, that is, falls within the influence field of a neuron in the decision space.

These types of networks work exactly like their software counterparts but without the excessive overhead required to run simulations. This makes them ideal candidates for SWAP constrained environments while offering many of the benefits of more experimental neuromorphic schemes. Cascaded networks exceeding 1 million parallel neurons are under development.

4.4 ASIC Large Scale Networks

Neuromorphic computing goals such as emulating a mouse's brain prove daunting due to the processing power required to emulate all 4 million neurons, much less that of a household cat, with approximately 300 million neurons. At the same time, extraordinary examples of pattern recognition and behavior are evident throughout the animal kingdom with significantly fewer neurons. For example, the round-worm, with 302 neurons and 8,000 synapses, can sense and track waterborne chemical signatures and navigate towards their locations [33]. Hardware neural networks built on ASICs have recently scaled beyond 40,000 parallel processors or neurons and are fast approaching million neuron networks. Since many typical classification problems can be addressed with as little as a few hundred active neurons, schemes for the optimization of larger networks need further refinement.

For example, such large scale parallel networks offer new possibilities for the speed-up of big data analysis. One such method to increase data throughput in parallel systems is to assign multiple processors the same operation but staggered in time using the single instruction multiple data (SIMD) scheme [34–36]. In this method, banks of identically trained neurons concurrently process high speed data with a linear reduction of processing times as 1/N, where N is the number of neuronal banks employed. The design is based on the single instruction multiple data (SIMD) paradigm as depicted in Fig. 8.

This approach takes advantage of the massively parallel nature of neuromorphic circuitry while utilizing surplus neurons for increased speed. Each neuronal bank is made from many individually trained processors that operate on the data independently. Within a given neuronal bank the multiple instructions, single data (MISD) paradigm is used to analyze the data. In other words, the data is fed to a

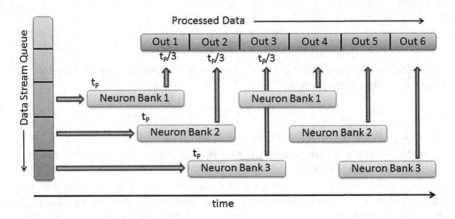

Fig. 8 A diagram depicting the SIMD process where multiple parallel neuron banks perform the same operations but are staggered in time resulting in increased data throughput where t_p is the processing time of a single neuron bank

common bus and is visible to all neurons in the bank for comparison to a target signature stored in the neuron's memory.

Another biologically inspired use of parallel processing is through multiple instruction, multiple data (MIMD) methods [37]. Here, every neuron, or small group of neurons, can operate on its own unique data stream to conduct specialized operations. This architecture is found in biological systems that are heavy on sensory perception where relatively weak processors act on numerous sensor inputs resulting in sensor-rich feedback controls. A fly for example, can have 80,000 sensory input sites and 338,000 neurons such that 98% of the neurons are used in perception [38]. The ratio of sensors to processers in these biological systems is in sharp contrast to many of our high performance systems that are built on few sensors with HPCs.

Lastly, serial architectures allow for the construction of decision trees and hierarchical networks where past computations affect current actions. Serial processing is, of course, fundamental to our perception and the traditional powerhouse behind computing performance. Only recently has the push for parallelism in computer architectures taken such a noteworthy role. The sequential transmission and analysis of sensory information within the brain is central, and in many senses is only complemented by parallel acceleration where appropriate.

Most designs will utilize multiple schemes within the overall system architecture. A simplified breakdown of basic processing methods discussed above is given below:

Single Instruction Multiple Data (SIMD)
Multiple processors performing the same operation but staggered in time
Increased data rates throughput parallelism
Banks of identically trained neurons
Utilize surplus neurons for increased speed

Multiple Instruction Single Data (MISD)
Many operations on same data at the same time
Functional parallelism
Each neuron in a bank has own unique target signature
Increased functionality and speed
Multiple Instruction Multiple Data (MIMD)
Many sensors processed simultaneously
Perceptual parallelism
Distributed memory
Biologically inspired
Serial Processing
Allows for hierarchical structures
Redundancy
Decision trees
Self-learning

5 Concluding Remarks

A brief look at the issues and challenges of developing solid state neuromorphic computing systems for SWAP constrained environments has been presented. Various concepts involving crossbar nanomatrices, CMOS/memristor hybrids, and ASIC artificial neural networks have been examined. Work continues to refine and mature the concepts so that they can be integrated into and utilized by future systems.

References

1. P.W. Singer, *Wired for War: The Robotics Revolution and Conflict in the 21st Century* (Penguin, New York, 2009)
2. W.J.A. Dahm (2012), Report on Technology horizons: a vision for air force science and technology during 2010–2030, http://www.aviationweek.com/media/pdf/Check6/USAF_Technology_Horizons_report.pdf. Accessed 04 Jan 2013
3. J. Misra, I. Saha, Artificial neural networks in hardware: a survey of two decades of progress, J. Neurocomput. **74**, 239–255 (2010)
4. D. Lammers (2012), Intel cancels Tejas, moves to dual-core designs. In EE Times. http://www.eetimes.com/electronics-news/4048847/Intel-cancels-Tejas-moves-to-dual-core-designs. Accessed 31 Jan 2013
5. M.D. Hill, M.R. Marty, Amdahl's law in the multicore era. Computer **41**(7), 33–38 (2008)
6. H. Esmaeilzadeh, E. Blem, R.S. Amant, K. Sankaralingam, D. Burger, Dark silicon and the end of multicore scaling, in *38th Annual International Symposium on Computer Architecture (ISCA)*, pp. 365–376, 4–8 June 2011

7. P.M. Kogge, Hardware Evolution Trends of Extreme Scale Computing, University of Notre Dame, 26 April 2011
8. J. Koomey, Growth in Data Center Electricity use 2005 to 2010 Analytics Press, Oakland (2011)
9. J. Blau (2012), Technology Time Machine 2012: Beyond CMOS, in *IEEE Spectrum*. http://spectrum.ieee.org/tech-talk/semiconductors/devices/technology-time-machine-2012-beyond-cmos?utm_source=feedburner&utm_medium=feed&utm_campaign=Feed%3A+IeeeSpectrum+(IEEE+Spectrum). Accessed 31 Jan 2013
10. Y.V. Pershin, M. Di Ventra, Adv. Phys. **60**, 145–227 (2011)
11. G. Snider, Prolog: Memristor Minds, in *Advances in Neuromorphic Memristor Science and Applications, Springer Series in Cognitive And Neural Systems 4*, ed. by R. Kozma, R. Pino, G. Pazienza (Springer, New York, 2012), pp. 3–7
12. H. Ames, M. Versace et al., Persuading Computers to Act More Like Brains, in *Advances in Neuromorphic Memristor Science and Applications, Springer Series in Cognitive and Neural Systems 4*, ed. by R. Kozma, R. Pino, G. Pazienza (Springer, New York, 2012), pp. 37–61
13. C. Yakopcic, T.M. Taha et al., Analysis of a memristor based 1T1M crossbar architecture, in *The 2011 International Joint Conference on Neural Networks (IJCNN), IEEE*, pp. 3243–3247 (2011)
14. D.B. Strukov, K.K. Likharev, Reconfigurable Nano-Crossbar Architectures, in *Nanoelectronics and Information Technology*, 3rd edn., ed. by R. Waser (Wiley, New York, 2012), pp. 543–562
15. GS. Rose, H. Manem et al., Leveraging memristive systems in the construction of digital logic circuits and architectures, in *Proceedings of the IEEE* , **100**(6), 2033–2049 (2012)
16. J. Rajendran, H. Manem et al., An energy-efficient memristive threshold logic circuit. IEEE. Trans. Comput. **61**(4), 6:1–6:22 (2012)
17. J. Rajendran, H. Manem et al., An approach to tolerate variations for memristor based applications, in *Proceedings of the 24th International Conference on VLSI Design (VLSI Design)* pp. 18–23 (2011)
18. J. Rajendran, H. Manem et al., Memristor based programmable threshold logic array, in *IEEE/ACM International Symposium on Nanoscale Architectures*, pp. 5–10 (2010)
19. J. Rajendran, R. Karri, G.S. Rose, Parallel memristors improve variation tolerance in memristive digital circuits, in *IEEE International Symposium on Circuits and Systems*. pp. 2241–2244 (2011)
20. H. Manem, G.S. Rose, Design considerations for variation tolerant multilevel cmos/nano memristor memory, in *ACM Great Lakes Symposium on VLSI*. pp. 287–292 (2010)
21. H. Manem, G.S. Rose, A Crosstalk Minimization technique for sublithographic programmable logic arrays, in *IEEE Conference on Nanotechnology*. pp. 218–222 (2009)
22. H. Manem, J. Rajendran, G.S. Rose, Design Considerations for Multi-Level CMOS/Nano Memristive Memory. ACM. J. Emerg. Technol. Comput. Syst. **8**(1), 1–22 (2012)
23. H. Manem, G.S. Rose, A read-monitored write circuit for 1T1 M memristor memories, *IEEE International Symposium on Circuits and Systems*. pp. 2938–2941 (2011)
24. M. Soltiz, C. Merkel et al., RRAM-based adaptive neural logic block for implementing non-linearly separable functions, in *Proceedings of IEEE/ACM International Symposium on Nanoscale Architectures (NANOARCH)* (2012)
25. M. Solti, D. Kudithipudi et al., Submitted 2012. Single-Layer Neural Logic Blocks Using Memristive Synapses, Submitted to IEEE Transaction on Computers (2012)
26. H. Manem, J. Rajendran, G.S. Rose, Stochastic gradient descent inspired training technique for a hybrid CMOS/Nano trainable threshold gate array. IEEE. Trans. Circuits. Syst. **59**(5), 1051–1060 (2012)
27. A.R. Omondi, J.C. Rajapakse, *FPGA Implementations of Neural Networks* (Springer, Netherlands, 2006)
28. A. Eide, T. Lindblad et al., An implementation of the zero instruction set computer (ZISC036) on a PC/ISA-bus card, in 1994 *WNN/FNN* (1994)

29. F.M. Dias, A. Antunes, A. Mota, Artificial Neural Networks: a Review of Commercial Hardware. Eng. Appl. Artif. Intell. IFAC **17**(8), 945–952 (2004)
30. The CogniMem Communique (2012) CogniMem Technologies, Inc., Folsom, 1(2). http://www.cognimem.com/_docs/Newsletters/CogniMem%20Communique',%20Vol%201,%20Issue%202.pdf. Accessed 31 Jan 2013
31. J.Y. Boulet, D. Louis et al., (1997) Patent US5621863, http://patft.uspto.gov/netacgi/nph-Parser?Sect2=PTO1&Sect2=HITOFF&p=1&u=/netahtml/PTO/search-bool.html&r=1&f=G&l=50&d=PALL&RefSrch=yes&Query=PN/5621863. Accessed 31 Jan 2013
32. Y. Q. Liu, D. Wei, N. Zhang, M.Z. Zhao Vehicle-license-plate recognition based on neural networks, in *Information and Automation (ICIA), 2011 IEEE International Conference on,* pp. 363–366 (2011)
33. K. Shen, C.I. Bargmann, The immunoglobin superfamily protein SYG-1 determines the location of specific synapses in C. Elegans. In Cell. **112**(5), 619–630 (2003)
34. K. Diefendorff, P.K. Dubey, How multimedia workloads will change processor design. In Computer **30**(9), 43–45 (1997)
35. R.E. Pino, G. Genello et al., Emerging neuromorphic computing architectures and enabling hardware for cognitive information processing applications. Air Force Research Lab Rome, Information Directorate (2010)
36. D. Shires, S.J. Park et al., Asymmetric core computing for US Army high-performance computing applications (No. ARL-TR-4788). Army Research Lab Aberdeen Proving Ground MD, Computational and Information Sciences Dir (2009)
37. B. Barney, Introduction to parallel computing. Lawrence. Livermore. Nat. Lab. **6**(13), 10 (2010)
38. R. Zbikowski, Fly like a fly [micro-air vehicle], in *Spectrum, IEEE* **42**(11), pp. 46–51 (2005)

Machine Learning Applied to Cyber Operations

Misty Blowers and Jonathan Williams

1 Introduction

Cyber attacks have evolved from operational to strategic events, with the aim to disrupt and influence strategic capability and assets, impede business operations, and target physical assets and mission critical information. With this emerging sophistication, current Intrusion Detection Systems (IDS) are also constantly evolving. As new viruses have emerged, the technologies used to detect them have also become more complex relying on sophisticated heuristics. Hosts and networks are constantly evolving with both security upgrades and topology changes. In addition, at most critical points of vulnerability, there are often vigilant humans in the loop.

Despite the sophistication of the current systems, there is still risk that there will be insufficient time and resources for the war fighter to respond in a contested environment. The cyber environment can change rapidly and evolving strategies are sometimes combined with sophisticated multi-stage phased attacks. Current automatic cyber offensive and defensive capabilities will not remain competitive when they heavily rely on human oversight, a set of pre-defined rules and heuristics, and/or threshold based alerting. An autonomous system is needed which can operate with some degree of self-governance and self-directed behavior. Machine learning can help achieve this goal. Machine learning techniques can help cyber analysts both defensively and offensively. This chapter will investigate machine learning techniques that are currently being research and are under investigation. The purpose of this chapter is to educate the reader on some machine learning methods that may prove helpful in cyber operations.

M. Blowers (✉) · J. Williams
Information Directorate, Air Force Research Laboratory, Rome, NY 13441, USA
e-mail: Misty.Blowers@rl.af.mil

J. Williams
e-mail: Jonathan.Williams@rl.af.mil

R. E. Pino (ed.), *Network Science and Cybersecurity*,
Advances in Information Security 55, DOI: 10.1007/978-1-4614-7597-2_10,
© Springer Science+Business Media New York 2014

2 Machine Learning

Machine learning is a branch of Artificial Intelligence that focuses on the study of methods and techniques for programming computers to learn. Learning is the science of getting computers to act without being explicitly programmed; it may be supervised, unsupervised, or semi-supervised, or it may be governed by the principals of Darwinian Evolution. However, the methods presented in this research may not be best suited for every challenge in cyber operations. A learning algorithm that performs exceptionally well in certain situations may perform comparably poorly in other situations [1, 2]. For some applications, the more simplistic methods are most appropriate.

2.1 Supervised Learning

In supervised learning, a "teacher" is available to indicate one of three things: whether a system is performing correctly, if it is achieving a desired response, or if it has minimized the amount of error in system performance. This is in contrast to the unsupervised learning where the learning must rely on guidance obtained heuristically [3, 4]. Supervised learning is a very popular technique for training artificial neural networks.

2.1.1 Artificial Neural Networks

The study of Artificial Neural Networks (ANNs) originally grew out of a desire to understand the function of the biological brain, and the relationship between the biological neuron and the artificial neuron. ANNs have become an increasingly popular tool to use for prediction, modeling and simulation, and system identification. Many sources in literature discuss the basic structure and implementation of ANNs [3].

ANNs consist of processing units, called neurons, or nodes, and the connections (called weights) between them. The ANNs are trained so that a particular input leads to a specific target output. A simplified illustration of this training mechanism is shown in Fig. 1. The network is adjusted based on a comparison of the output and the target until the network output matches, or nearly matches, the target [5].

ANNs have the ability to derive meaning from complicated or imprecise data. They can be used to extract patterns and detect trends that are too complex to be noticed by either humans or other computer techniques.

In order to consider the operation of ANNs it is important to introduce some of the terms used. The neuron forms the node at which connections with other neurons in the network occur. Unlike the biological neural networks which are not

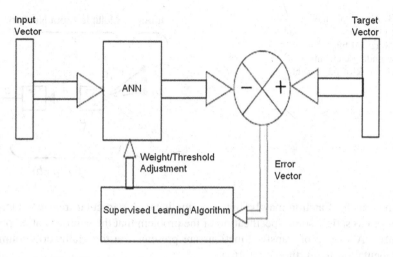

Fig. 1 A simplified illustration of the neural network training mechanism. The network is adjusted based on a comparison of the output and the target until the network output matches, or nearly matches, the target

arranged in any consistent geometric pattern, those in the electronic neural network are generally arranged in one or more layers which contain neurons performing a similar function. Depending on the type of network, connections may or may not exist between neurons within the layer in which they are located.

A single-input neuron is shown in Fig. 2. The scalar input, p, is multiplied by the scalar weight, w to form wp, which is one of the terms that is sent to the summer. If the neuron includes a bias, another input, 1, is multiplied by a bias, b, and then it is passed to the summer. The summer output, n, often referred to as the net input, goes into a transfer function, f, which produces the scalar neuron output, a.

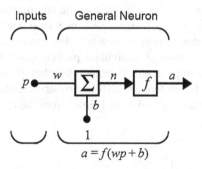

$$a = f(wp + b)$$

Fig. 2 Within a single input neuron the scalar input, p, is multiplied by the scalar weight, w, to form wp, which is one of the terms that is sent to the summer. If the neuron includes a bias then it is also passed to the summer. The net input, n, goes to the transfer function, f, which produces the scalar neuron output, a

Fig. 3 A multiple input neuron inputs p_1, p_2, ...pr are each weighted by corresponding elements w_1,1, $w_{1,2}$,....w_1, R of the weight matrix W

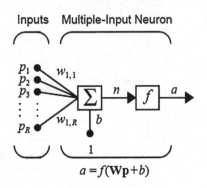

$$a = f(\mathbf{Wp} + b)$$

The transfer function may be linear or nonlinear. A particular transfer function is chosen to satisfy some specification of the problem that the neuron is attempting to solve. A variety of transfer functions are presented in the Mathworks training documentation for further study [6].

Typically, a neuron has more than one input. A neuron with multiple inputs is shown in Fig. 3. The individual *inputs* p_1, p_2, ..., p_r are each weighted by corresponding elements $w_{1,1}$, $w_{1,2}$,..., $w_{1,R}$ of the weight matrix W.

Some neural network models have several layers of networks. Each layer has its own weight matrix, its own bias vector, and net input vectors. Internal layers are often hidden to simplify the model. Hidden layers within the network also take part in producing output when the training is complete. The number of hidden layers is problem dependent, as an increase in the number of hidden layers increases the complexity [5].

It is important, however, to recognize the limitations of ANNs. ANNs have generally been shown to perform as very good multi-dimensional interpolators. In this context, they are limited by the boundaries of the information submitted to them during the training phase of their development. For real world applications, if the bounds of the information provided during the training phase does not extend to cover the entire region of anticipated future interest, then the network model must be retrained when changes are made to the environment that they model.

In an Intrusion Detection System, an ANN may be trained to recognize patterns characteristic of malicious activity or an attack. For example, as an input, the ANN could receive all the features an individual packet. Then, based on the weights established during training, it would output whether it believed the packet to represent an intrusion. Given that an ANN is a supervised learner, it would require a labeled data set for training, and would suffer from the drawbacks mentioned in the preceding paragraph.

2.2 Unsupervised Learning

In machine learning, unsupervised learning is a class of problems in which one seeks to determine how the data are organized. It is distinguished from supervised learning and reinforcement learning in that the learner is given only unlabeled examples [7].

Barlow [8] explains the biological parallel of unsupervised learning, and how these algorithms provide insights into the development of the cerebral cortex and implicit learning in humans.

According to Barlow, much of the information that pours into our brains arrives without any obvious relationship to reinforce and is unaccompanied by any other form of deliberate instruction. The redundancy contained in these messages enables the brain to build up its "working modules" of the world around it. Redundancy is the part of our sensory experience that distinguishes meaningful information from noise [9]. The knowledge that redundancy gives us about patterns and regularities in sensory information is what drives unsupervised learning. With this in mind, one can begin to classify the forms that redundancy takes and the methods of handling it [8].

During the analysis of any large dataset, as is the case in cyber operations, the researcher needs assistance in finding the relevant data. In order to find patterns out of what appears to be chaos, clustering can be used. Cluster analysis is used to classify samples automatically into a number of groups using measures of association [10]. It can help the researcher find hidden relationships, allowing them to have a better understanding of the data and to build models that capture associated patterns of behavior.

2.2.1 Cluster Analysis

Two major categorizations of cluster analysis methods are discussed in this chapter: density-based and partition based. The key differences between them are in the way the clusters are formed.

Density-based approaches apply a local cluster criterion. Clusters are regarded as regions in the data space in which the objects are dense, and which are separated by regions of low object density (noise). These regions may have an arbitrary shape and the points inside a region may be arbitrarily distributed. DBSCAN is an example of this method. This will be explained in more detail later in this chapter.

In partitioning clustering, the data set is initially partitioned, often randomly, into k clusters. These initial clusters are then iteratively refined using some method until the chosen partitioning criterion is optimized. The K-Means algorithm is one of the most well-known and commonly used partitioning methods.

The K-means algorithm is a method of cluster analysis that aims to partition m observations (samples) into k clusters in which each observation belongs to the cluster with the nearest mean. The basic algorithm depicted in algorithm 1 [11, 12].

Algorithm 1 K-Means Algorithm

1. *Begin initialize m, c, μ_1, μ_2, ..., μ_c*
2. *do classify n observations according to nearest μ_i*
3. *recompute μ_i*
4. *until no change in μ_i*
5. *return μ_1, μ_2, ..., μ_c*
6. *end*
7. *where m is the number of observations, c is the number of clusters, and u_i is a specific cluster centroid within the set of clusters from 1 to c.*

One criticism in using the K-Means algorithm is that the number of groups must be predefined before creating the clusters. In other words, it is sometimes difficult to know how to pick the best value for K. Choosing a number smaller than the number of naturally occurring clusters yields centroids that include many unrelated samples. The selection of a number larger than the number of naturally occurring clusters yields centroids that are competing for highly related samples. To overcome this obstacle, a cluster evaluator may be used which attempts to minimize the inter- to intra-cluster distance.

Figure 4 provides further clarification on the K-Means algorithm [13]. In the first step, each point is assigned to one of the initial centroids to which it is closest, depicted as a cross. After all the points are assigned to a centroid, the centroids are updated. In step two, points are assigned to the updated centroids, and then the centroids are recalculated again. In Fig. 4, steps 2, 3, and 4 show two of the centroids moving to the two small groups of points in the bottom of the figures. The algorithm terminates in step 4, because no more changes need to be made to the centroids. The centroids have identified the natural grouping of clusters [13].

To assign an observation to a closest centroid, as described in the preceding paragraph, a proximity measure is required to quantify the notion of "closest". Most often, one of two distance measures are used: the Euclidean (L2) distance measure and the Manhattan (L1) distance [13]. While the Euclidean distance

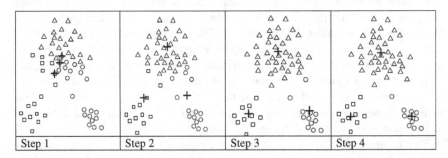

| Step 1 | Step 2 | Step 3 | Step 4 |

Fig. 4 The K-means algorithm is an example of a partitioning method. This illustration shows how it can be used to find three clusters in a sample of data

corresponds to the length of the shortest path between two samples (i.e. "as the crow flies"), the Manhattan distance refers to the sum of distances along each dimension (i.e. "walking round the block"). The Euclidean distance d_E is defined by Eq. 1, and the Manhattan distance d_M (or city-block distance) is defined by Eq. 2 [10].

Given N observations, $x_i = (x_{i1}, ..., x_{in})^T$, $i = 1, 2, ..., N$ the following equations may be used to find the distance between the jth and kth observations.

$$d_E(j,k) = \sqrt{\left(\sum_{i=1}^{n} |x_{ji} - x_{ki}| \wedge 2 \right)} \tag{1}$$

$$d_M(j,k) = \sum_{i=1}^{n} |x_{ji} - x_{ki}| \tag{2}$$

There have been several works on calculating distance to solve certain pattern recognition problems [14–16]. The methods used depend on the nature, and size, of the data. It also depends on if the algorithm is being used with the k-means clustering method, or another method. Experiments were conducted with both distance functions to see which would perform the best.

In addition to the previously noted challenge of picking the right number of starting clusters, another common criticism of the K-Means algorithm is that it does not yield the same result with each run. This is attributed to the fact that the resulting clusters depend on the initial random assignments [17]. This obstacle can be overcome, however, by fixing the initial assignments. In the implementation of the K-Means algorithm, the random number generator used for initial assignments was provided the same seed, resulting in repeatable pseudo-random numbers. Still another disadvantage is that when the data set contains many outliers, K-Means may not create an optimal grouping. This is because the outliers are assigned to many of the allocated groups. The remaining data becomes divided across a smaller number of groups, compromising the quality of clustering for these observations [18].

One extension of the K-Means algorithm is the quality threshold k-means. Quality is ensured by finding clusters whose diameter does not exceed a given user-defined diameter threshold. This method prevents dissimilar samples from being forced under the same cluster and ensures that only good quality clusters will be formed [19].

Another extension of the K-Means algorithm is the fuzzy-k-means algorithm. This algorithm is based on concepts from fuzzy set theory. Fuzzy set theory is based on the idea that humans work with groups of entities that are loosely defined, able to admit elements according to some scale of membership rather than an absolute yes/no test [20]. In using fuzzy k-means, a sample's membership to a cluster is a function of its distance to the centroid. While it does not solve any of the initialization problems of k-means, it offers users a soft degradation of membership instead of the hard binary relations of the original algorithm.

K-means is excellent for partition-based cluster analysis but DBSCAN is useful when a density-based analysis solution is needed. DBSCAN stands for density based spatial clustering of applications with noise. It focuses on the notion of density reachability [21]. A point q is directly density reachable from a point p if q is within a set distance, epsilon (ε), from p, and there is also a set number of points, called "minPts", within a distance of q from p. Two points, p and q, are density reachable if there is a series of points from p to q that are all directly density reachable from the previous point in the series. All points within a cluster formed by DBSCAN are density reachable, and any point that is density reachable from any point in the cluster, belongs to the cluster. The basic DBSCAN algorithm operates as follows:

Algorithm 2 DBSCAN

1. *Choose an unvisited point P from the data set and mark it as visited. Finished if there are none.*
2. *Find all the neighbors of point P (points within a distance of (ε).*
3. *If the number of neighbors is less than minPts, mark P as noise and return to step 1.*
4. *Create a new cluster and add P to the cluster.*
5. *Repeat steps 6-9 for each point P' of the neighbors of P*
6. *If P' is not visited, mark it as visited. Otherwise return to step 5.*
7. *Find the ε neighbors of P'*
8. *If the number of neighbors of P' is not less than minPts, join the neighbors of P' with the neighbors of P.*
9. *If P' does not belong to a cluster, add it to the cluster of P.*
10. *Return to step 1*

Much like the K-Means clustering algorithm, DBSCAN has the issue of choosing the proper inputs to the algorithm. DBSCAN takes two inputs, epsilon and minPts. In an attempt to find the optimal value of ε, a ε evaluator may be used. The evaluator relies on the use of a k-distance graph. To generate this graph, the distance to the kth nearest point is determined for each point that is to be clustered. Generally, minPts is used as the value for k. This list of distances is then sorted from smallest to largest. The plotting of this list results in a gradually upward sloping line that suddenly increases greatly in slope towards the end. Choosing epsilon to be the value of the distance just before the large increase in slope, results in the best clustering. Noise points whose k-nearest distance is large will not be clustered, and core points with a small k-nearest distance will be clustered with nearby points (Fig. 5).

Fig. 5 In the first image, points p and q are directly density reachable. In the second image, points p and q are density reachable [22]

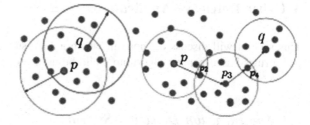

2.3 *Evolutionary Computation and Genetic Algorithms*

Another method in machine learning is through evolutionary computation and genetic algorithms. The Genetic Algorithm is based on an analogy to biological evolution. An initial population is created consisting of a learned set of generated rules. Based on the notion of survival of the fittest, a new population is formed to consist of the fittest rules and their offspring. The fitness of a rule is represented by its classification accuracy on a set of training examples. Offspring are generated by crossover and mutation. The process continues until a population P evolves when each rule in P satisfies a pre-specified threshold. This method is sometimes slow but easily parallelizable. Seeded solutions may also be injected into the solution set to speed up the computation speed.

Figure 6 shows a simplistic example of how the genetic operators work.

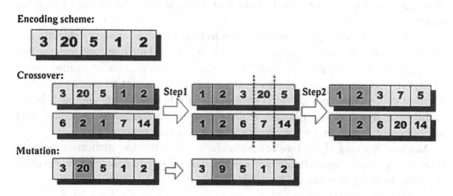

Fig. 6 The figure shows how the genetic operators work. With a single point cross over one crossover point is selected, then this portion of the string is copied from the beginning of the chromosome to the crossover point, the remainder is copied from the other parent to create a new chromosome. Mutation simply changes a single bit on the chromosome

3 Cyber Defensive Applications

This section will discuss methods in which a semi-supervised learning approach and a rules-based genetic algorithm could be used for intrusion detection.

3.1 The Intrusion Detection System

Security concerns are becoming increasingly important to the Department of Defense with the increasing reliance on networking capabilities to provide quick and reliable information to the analyst. It is estimated that the ratio of false alarms to true alarms in most commercial signal based intrusion detection systems may vary from 5:1 to 20:1. Too many false alarms overwhelm the system operator and eventually the alarms become ignored [23]. Another challenge in intrusion detection is the fact that normal behavior continuously changes and new attacks continuously emerge. The environment is dynamic and there is an increasing need for robust solutions.

For this reason, an Adaptive Cluster Tuning (ACT) method using semi-supervised learning approach for targeted detection and characterization of cyberattacks was investigated [24]. This approach controls the number of alarms output to the system operator, and allows the operator to tune the detection model based upon the emergence of patterns of events that deviate from normal behavior. It also allows the analyst to focus in on more threats that are important and relax the bounds around threats that are of lower priority. With such a system the analyst may have access to information about that detailed characterization of a given attack.

There are two types of major detections to be considered; misuse detection and anomaly detection. Attacks are one type of anomaly and may be further described in terms of attacks against availability (denial of service), attacks against confidentiality (eavesdrop), and attacks against integrity (attempt to modify communication contents or data in the system). One key benefit of the approach is the ability to assign attack attribution to the events of interest based upon historical attribution of events that were characterized by similar patterns.

Machine learning algorithms are particularly useful for this application because there are valuable implicit regularities and anomalies that may be discovered. Machine learning also allows for adaptation to changing conditions.

3.1.1 Data Analysis

Although this research is not limited to a single data set, the data set that was initially evaluated is the data set used for The Third International Knowledge Discovery and Data Mining Tools Competition, which was held in conjunction with KDD-99, The Fifth International Conference on Knowledge Discovery and Data Mining. The competition task was to build a network intrusion detector, a predictive model capable of distinguishing between intrusions or attacks, and normal conditions [25]. This database contains a standard set of data including a wide variety of intrusions simulated in a military network environment. This data set is intended to be used as a starting point to build the framework for the proposed approach. It is fully understood that new threats have emerged since this data set was developed, and it is the intention that this program will investigate data sets that are more recent as they are identified and become available.

The full KDD CUP training dataset contains 4,898,431 connection records. A connection is a sequence of TCP packets starting and ending at some well-defined times, between which data flows to and from a source IP address to a target IP address under some well-defined protocol. Associated with each connection record are 41 different features. These features are a mixture of continuous and symbolic data types.

A file containing 10 % of the training data contains 494,021 connection records. There are 97,278 normal records and the other 396,743 records are various types of attacks. In addition, there is an unlabeled test data set containing 2,984,154 connection records, and a file containing 10 % of the test data with 311,029 records. There is also a labeled version of the 10 % of the test data. In that file, 60,593 of the records are "normal" and the other 250,436 records are various types of attacks. The file containing 10 % of the training data contains 22 different attack types and the file containing 10 % of the test data contains 38 different attack types.

Each connection record is labeled either as "normal" or as a specific attack type. There are four main categories of attack types; DOS (denial of service), R2L (unauthorized access from a remote machine), U2R (unauthorized access to a local super user privileges, and probing (surveillance and other probing).

The datasets lend themselves very well to existing semi-supervised learning algorithms that have been applied to other research domains by the PI. Each connection record is a discrete state that can be represented as a feature vector and is labeled as either normal or an attack. The specific attack type is specified, and this information will be used for the attribution study of the proposed approach. Nine of the features for each record are symbolic or discrete which are not generally useful in the current algorithmic, but this still leaves 32 useful features.

3.1.2 Data Pre-Processing

The first part of the ACT system is to filter out the poor quality data. Care needs to be taken to ensure that we do not remove anomalies that may be characteristic of events of interest. Since the data set under consideration is a simulated data set, this step may not initially seem as important as it is when a real world data set is considered. However, this step is especially important when trying to discriminate

attacks that periodically generate false alarms because they are so closely related
to events that may be normal.

One of the simplest of the feature selection processes is an algorithm that was
developed to make use of some very simple statistical analysis methods [24, 26].
This method helps prune the data set so that the most relevant features are used for
the analysis. The Tukey Quick Test is a simple statistical test that has been used to
find which features vary the most between two classes. In prior work [24], this
principal was demonstrated to be useful because of its speed and robustness when
dealing with large data sets, and subsequent real time processing. The entire data
set is considered as the total population and each class is considered a sample from
the population. The basic algorithm is depicted in Algorithm 3 [26].

Algorithm 3 Feature Selection Algorithm

1. *Locate the largest and smallest values of the overall population.*
2. *If both the largest and smallest value occurs in the same sample, we cannot
   conclude that the means of the two samples are different.*
3. *Locate the largest and smallest values in sample 1.*
4. *Locate the largest and smallest values in sample 2.*
5. *Consider the sample containing the smallest value in the two samples combined, as found in Step1. Count the number of values in this sample that are
   smaller than the smallest value in the other sample.*
6. *Consider the sample containing the largest value in the two samples combined,
   as found in Step 1. Count the number of values in this sample that are larger
   than the largest value in the other sample.*
7. *The sum of the counts obtained in Step 5 and 6 is the numerical value of the test
   statistic T.*

Correlation methods like Spearman's rank correlation will provide further
analysis of the features selected from the process described above. Spearman's
rank correlation is a nonparametric (nonlinear) correlation analysis that may be
used to determine the correlation of each proposed input variable to the output
variable (X_i, Y_i). The analysis is a measure of the strength of the associations
between two variables of a collection of subsets of data. The total number, n, raw
scores (X_i, Y_i) are converted to ranks x_i, y_i. The differences $d_i = x_i - y_i$ between
the ranks of each observation on the two variables are calculated. The rank correlation is then determined by the formula described in Eq. 3.

$$\rho = 1 - \frac{6 \sum d_i^2}{n(n^2 - 1)}.$$ (3)

Because the Spearman rank correlation coefficient uses ranks, it is much easier
to compute than other correlation methods. Once the rankings have been calculated, a threshold value is selected which will determine if the variables identified
are significantly correlated enough to be retained in the set of input variables.
Those variables not significantly correlated are not selected.

Another statistical measurement that may be used to rank features is information gain. Information gain measures the reduction of entropy achieved by learning the value of a particular feature. A more detailed explanation on information gain can be found in the following Ref. [27].

3.1.3 Semi-Automated Intelligence with Analyst Feedback

As previously noted, in developing intelligent IDS some consideration must be given to the desired level of autonomous operation. A recent US Patent Application [28] shows a method that allows for adjustment of false alarms for objects or events of higher interest with a semi-supervised learning approach. This approach controls the number of alarms output to the system operator, and allows the operator to tune the detection model based upon the emergence of patterns of events that deviate from normal behavior. It also allows the analyst to focus on threats that are more important and relax the bounds around threats that are of lower priority. A graphical user interface may be used to provide the analyst access to information about that detailed characterization of any given attack.

The clustering methods described in this chapter have an added benefit in that they enhance the class separation to help discriminate one attack type from another and to discriminate periods of normalcy. This is important because the attacker sometimes tries to disguise the attack to make things appear normal.

In this example, major classes to be considered will be a class of normalcy and the classes of system attack. Within each class, several sub-classes will emerge. The main classes are used for the supervised portion of the learning system. The sub-classes that emerge do so in a manner that would be considered "unsupervised", or without prior example. Because many "normal" operations appear similar to periods of times when the system is under attack, there is a great deal of overlap between the two classes. For this reason, each class is characterized separately. Figure 7 helps illustrate this point.

Within this method, the model can be adjusted by changing threshold values to accommodate the tolerance level for false alarms or the cost benefit from early detection. In order to understand what information the various clusters represented, the information from the attack categories attack types are mapped to the clusters which were defined by attack indicator vectors. Further research and analysis is needed to verify that the clusters formed are unique to certain attacks of interest.

In cases where there is significant overlap in the feature space, the clusters formed are considered "weak clusters". For these regions of the feature space where significant overlap exist, the clusters may be eliminated or adjusted based upon the analyst feedback.

Normalcy is also characterized by clusters. The input vectors that vary significantly from both the previously characterized events and the clusters of normalcy may be flagged, on the fly, and the cluster they form may prompt a response from the end user to characterize the unknown as a new event of interest or a state of normalcy. It will also prompt the user to enter the suspected causes of the event.

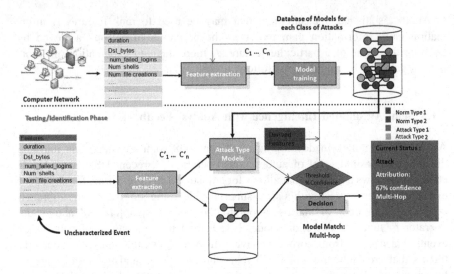

Fig. 7 Example of machine learning process for intrusion detection. The *top row* shows the learning phase where a repository of models is developed for each class and subclass of attack types. In the testing and identification phase, an uncharacterized event will be compared to the trained models

The system can then be updated so that when these new events are encountered in the future, they will be identified.

The steps described in Fig. 8 are further detailed below:

1. Data preprocessing
   The first step in this method is the data pre-processing step. In operation, an analyst would load the data file and select the interval that will serve as the training data set. For extreme large data sets, it is important that the analyst use a feature selection technique to determine which features to use to create the feature vectors that will be mapped into the multi-dimensional feature space.

   Feature analysis table: Various statistical analysis techniques may be performed on the data that to determine the optimal inputs. These results should be graphically displayed for the analyst so that they may have the option to manually override the features which will be used an inputs to the model. An example of this graphical interface is shown in Fig. 9.

2. Clustering/threshold adjustment
   Training: The next step is to associate the designated feature vectors into their clusters and evaluate/optimize the thresholds.

   Threshold optimization: After the clusters have been trained, the thresholds for each of the clusters are optimized. To do this, first, each data entry from the validation set is assigned to the nearest cluster. The validation set consists of data

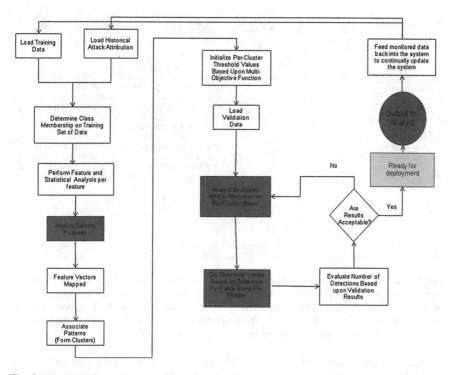

Fig. 8 The figure is an example of how an adaptable cluster tuning (ACT) method may be used for an IDS. The analyst could have input into the system as described above. Analyst inputs and outputs are shown in *red*

from the file initially loaded in the data-preprocessing screen, beginning where the training set ended and spanning the same amount of time as the training set. The thresholds for each cluster are then dynamically optimized in an attempt to maximize the detection rate accuracy while also meeting the selected minimum detection rate. The detection rate is the percentage of attack indicating entries in the validation set that are correctly classified as attack-indicating (fall within their cluster's threshold). The false alarm rate is the percentage of non-fault indicating records in the validation set which are incorrectly classified as attack-indicating.

3. Event attribution: At this stage of the process, the analyst may view the attack types associated with the various patterns that have formed. Since more than one event may be associated with a given cluster, this report may be most useful if it is given as a breakdown of percentage of associated attacks with each cluster.

   Desired detection slider: The detection rate is the percentage of correct detections in the validation set that are correctly classified as detections. A desired detection slider allows the user to select the minimum detection rate on a per cluster basis.

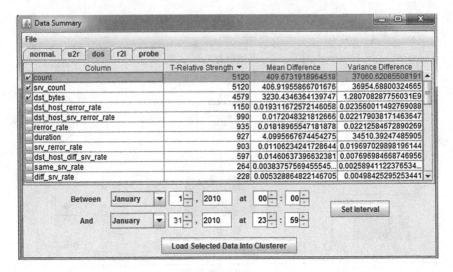

Fig. 9 An example interface for feature extraction is shown above. Each *row* in the table represents a column (or particular feature) from the original data set. In this example, the first value in the row is the name of the data feature. The next three values are the T-relative strength, information gain and mean difference for the data from that sensor. Each column in the table can be sorted by clicking on the column header

4. Start intrusion monitoring system: The method is now configured so that an analyst or operator can evaluate future attacks of interest. An analyst is notified when an attack is present and provided with information about the attack attribution in the form of a pie chart. Feedback is provided to the system based upon the operator's inputs.

Figure 10 shows the performance of the Quality Threshold K-Means Method as compared to some other machine learning approaches. It should be noted that this method has the added advantage over the rest in that it allows for analyst adjustment for key targets of interest. These values have been optimized for comparison purposes, but the user should note that adjusting one region of the feature space for high priority attack types might cause the overall performance of this method to decrease. The reason for this is because there will be lower reward for detecting attack types of lower priority.

4 Genetic Algorithm Applications

Genetic algorithms have many uses in the cyber domain. A Genetic Algorithm (GA) is a family of computational models based on principles of evolution and natural selection. A GA employs a set of adaptive processes that mimic the concept of "survival of the fittest". They have been studied in a robust manner [29, 30] and with specific applications in mind.

Fig. 10 The figure shows the performance of different machine learning methods on the KDD data set. The average percentage accuracy is determined by the number of correctly classified attack types and the correctly classified periods of normal operation

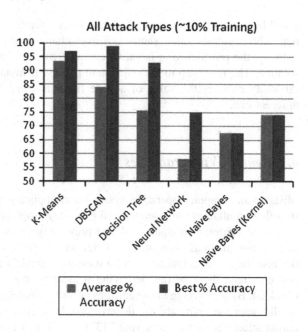

Chief among these applications is unbiased, real-time detection and analysis of anomalies across an entire network [31]. Genetic algorithms may provide a significant advantage to both defender and attacker, but the advantage is far stronger for the defender because anomalies can be detected within digital traffic on a computer network [32]. Some IDS use genetic algorithms [33–35] to train their detectors while other IDS are entirely based on these concepts [36]. They create an environment in which the attacker would need to camouflage network traffic signatures in addition to the already difficult tasks of finding and exploiting vulnerabilities. A proactive defender can therefore require an attacker to add tremendous complexity, time and cost to their tactics to bypass defenses.

Defender's systems will harden with each attack because machine learning becomes more intelligent and effective over time thus creating exponentially more secure systems over time [30]. This technology has the potential to enable cyber security systems to detect, evaluate and counter threats by assessing anomalies within packets, data traffic sessions and user behaviors across the entire network [31]. The eventual result of network defended by these techniques will be trustworthy enclaves in cyberspace that will have observable security metrics that may be modeled so that abnormalities are readily identified and acted upon [29].

The genetic algorithm approach starts with an initial population (often times randomly generated), and then evolves optimal solutions through selection, crossover, and mutation. Finally, the best individual is selected once the optimization criterion is met [38, 39, 40]. The GA evolves the existing set of chromosomes by combining and refining the genes contained within each chromosome. The objective is to produce new chromosomes that form a new generation of possible solutions to evaluate. The crossover operation allows the GA to create new chromosomes that share positive characteristics while simultaneously reducing the prevalence of negative characteristics in an otherwise reasonably fit solution. The final step in the refinement process is mutation. The mutation phase randomly changes the value of a gene from its current setting to a completely different one.

4.1 Remote Vulnerabilities

Mistakes in computer operating systems and applications may create vulnerabilities that an attacker can identify and exploit. A sophisticated attacker might use genetic algorithm techniques to detect previously unknown vulnerabilities. However, these methods also provide defenders with an advantage over attackers because the same techniques can be used by defenders to identify vulnerabilities and eliminate them before deploying a system as a part of a routine security checklist. By identifying vulnerabilities before an attacker does, defenders are able to eliminate the vulnerability, deceive attackers by masking the vulnerability or entrap attackers with a "honeypot" [37] system that is not truly vulnerable [30].

Vulnerabilities can be eliminated by patching software or reducing and/or eliminating the attack surface. Software patches are routine computer administrative tasks. Attack surface reduction can be achieved by configuring a firewall to block network access to a vulnerable resource or by taking a vulnerable system offline to prevent the vulnerability from being used to cause further damage. Deception can be achieved by redirecting logical network addresses to protected areas. Entrapment is a more robust type of deception usually achieved with a honeypot. These methods can be greatly improved by using a genetic algorithm.

4.1.1 Hybrid Honeypot

The Hybrid Honeypot method makes use of genetic algorithms to improve the defenders objectives. Honeypots are decoy computer resources set up for the purpose of monitoring and logging the activities of entities that probe, attack or compromise them [37]. A honeypot is useful for deceiving an attacker and causing them to spend valuable resources (primarily time) within the decoy environment. The attacker may think that they are delving deeper into a vulnerable network, but they are actually attacking an environment setup specifically to entrap and monitor their activity.

The hybrid honeypot used a network tap, span port, hub, or firewall, and collected traffic based information that traversed a given network. This traffic data was then used by the GA for the creation of a set of rules for an Intrusion Prevention Rule based system. Intrusion prevention follows the same process of gathering and identifying data and behavior, with the added ability to block or prevent the activity [37].

The hybrid architecture combined the best features of low and high interaction honeypots. It helped in detecting intrusions attacking the system. The low interaction honeypot provided enough interaction to attackers to allow honeypot to detect interesting attacks. It provided information about attacks and attack patterns [37].

5 Machine Learning for Behavior Analysis for Insider Threats

Machine learning once again provides the defender more advantages than the attacker when considered in the case of insider threats. Although attackers can use machine learning to impersonate the "signature" activities of an identity, they are limited to behaving exactly as that identity would [30]. The attacker must know how that person (identity) has behaved in the past and how the system will perceive their every movement, especially from a security and permissions standpoint [30].

The defender's advantage is that machine learning creates a "pattern of life" for every known, authenticated user. This representation of all past behavior can be monitored both on a particular workstation and across a computer network. A robust pattern of life for all known users allows network defenders to evaluate patterns and identify anomalies with a degree of precision that would be very difficult to achieve with conventional computer programming techniques [30].

6 Conclusion

Machine learning is at the forefront of network defense technology but humans will always have a role in the decision making process. It can enable humans to make rapid, well-informed decisions through network-wide sensing coupled with unbiased detection and analysis [29].

Machine learning techniques present methods for making sense of cyber data, alerting when suspicious and possibly malicious activity occurs on monitored networks, and provides a means to design strategic cyber operations which carry out the intended effects rapidly and with greater precision. There is growing complexity associated with today's complex network environments and there is not enough man-power to manually analyze the increasing amounts of data. It is becoming more and more difficult to understand cyber operations and vulnerabilities. The adaptive and autonomous techniques discussed in this chapter show promising results to be of great use to the cyber warriors of tomorrow.

References

1. D.H. Wolpert, The lack of a priori distinctions between learning algorithms. Neural Comput **8**(7), 1341–1390 (1996)
2. H Zhu. No free lunch for cross validation. Neural Comput. **8**(7), 1421–1426 (1996)
3. K. Mehrotra, C. Mohan, S. Ranka, *Elements of Artificial Neural Networks* (The MIT Press, Cambridge, 2000)
4. E. Baum, F. Wilczek. Supervised learning of probability distributions by neural networks. Neural Inf. Proces. Syst. **1** 52–61 (1988)
5. G. Scott. *Knowledge-based artificial neural networks for process modeling and control* (The University of Wisconsin, Madison, 1993)
6. Documentation-Neural Network Toolbox. Mathworks. [Online] [Cited: January 13, 2010.] http://www.mathworks.com/access/helpdesk/help/toolbox/nnet/neuron_2.html#33674
7. J. Woodward, Computable and incomputable functions and search algorithms. Intell. Comput. Intell. Syst. **1**, 871–875 (2009)
8. H.B. Barlow, Unsupervised learning. Neural Comput **1**(3), 295–311 (1989)
9. H. Barlow. *Possible principles underlying the transformations of sensory messages.* ed. by W. Rosenblith. Sensory Communication (MIT Press, Cambridge, 1961)
10. Z. Li, J. Yuan, H. Yang, K. Khang. K-mean Algorithm with a distance based on characteristics of differences, in *4th International Conference on Wireless Communications, Networking and Mobile Computing*, Oct 2008, pp. 1–4
11. J. Han, M. Kamber, *Data Mining: Concepts and Techniques* (Academic Press, San Diego, 2001)
12. R. Duda, P. Hart, D. Stork, *Pattern Classification* (John Wiley & Sons, New York, 2001)
13. T. Pang-Ning, M. Steinbach, V. Kumar. *Introduction to Data Mining* (Addison-Wesley, Boston, 2005)
14. J. Yu, J. Amores, N. Sebe, P. Radeva, Q. Tian, Distance learning for similarity estimation. IEEE Trans. Pattern Anal. Mach. Intell. **30**(3), 451–462 (2008)
15. C. Domeniconi, J. Peng, D. Gunopulos, Locally adaptive metric nearest neighbor classification. IEEE Trans. Pattern. Anal. Mach. Intell. **24**(9), 1281–1285 (2002)
16. E. Xing, A. Ng, M. Jordan, S. Russell. Distance metric learning, with application to clustering with side-information. in *Proceedings of Neural Information Processing Systems*, 2003, pp. 505–512
17. J. MacQueen. Cluster analysis. wikipedia. [Online] http://en.wikipedia.org/wiki/Cluster_analysis#cite_note-0
18. G. Myatt. *Making Sense of Data: A Practical Guide to Exploratory Data Analysis and Data Mining* (John Wiley & Sons, New York, 2007
19. L. Heyer et al. Exploring expression data: identification and analysis of coexpressed genes. Genome Res. **9**, 1106–1115 (1999)
20. L. Zadeh. Fuzzy sets. Inf. Control **8**, 338–353 (1965). As sited by Martin
21. M. Ester, H.-P. Kriegel, J. Sander, X. Xu. A density-based algorithm for discovering clusters in large spatial databases with noise, in *Proceedings of the Second International Conference on Knowledge Discovery and Data Mining (KDD-96)*, ed. by E. Simoudis, J. Han, U.M. Fayyad. AAAI Press, pp. 226–231. ISBN 1-57735-004-9
22. Pervasive and Mobile Computing [Online] http://www.sciencedirect.com/science/article/pii/S1574119209000686
23. H. Gunes Kayacik, A.N. Zincir-Heywood, M.I. Heywood, S. Burschka. Generating mimicry attacks using genetic programming: a benchmarking study, in *IEEE Symposium on Computational Intelligence in Cyber Security*, 2009
24. M. Blowers. Analysis of Machine Learning Models and Prediction Tools for Paper Machine Systems. Dissertation, State University of New York College of Environmental Science and Forestry, 2010
25. KDD Cup 1999 Data [Online] http://kdd.ics.uci.edu/databases/kddcup99/kddcup99.html

26. W. Stitieler, *Multivariate Statistics with Applications in Statistical Ecology* (International Co-operative Publishing House, Fairland, 1979)
27. J. Han, M. Kamber. *Data Mining Concepts and Techniques,* 2nd edn (Morgan Kaufmann, San Francisco, 2006). ISBN 1-55860-9016
28. M. Blowers, C. Salisbury. Method and apparatus for event detection permitting per event adjustment of false alarm rate. United States Patent Application 20120271782
29. G.A. Fink, C. S. Oehmen. Final Report for Bio-Inspired Approaches to Moving-Target Defense Strategies. No. PNNL-21854. Pacific Northwest National Laboratory (PNNL), Richland, WA , 2012
30. The Current State of Cyber Security is Fundamentally Flawed [Online] http://www.ai-one.com/2011/10/18/the-current-state-of-cyber-security-is-fundamentally-flawed/
31. Machines Can Learn the Inherent Complexity of Data [Online] http://www.ai-one.com/2011/10/17/machines-can-learn-the-inherent-complexity-of-data/
32. S. Borg. Securing the Supply Chain for Electronic Equipment: A Strategy and Framework. (2009) http://www.whitehouse.gov/files/documents/cyber/ISA%20-%20Securing%20the%20Supply%20Chain%20for%20Electronic%20Equipment.pdf
33. R.M. Chandrasekaran, M. Govindarajan. Bagged genetic algorithm for network intrusion detection. Int. J. Res. Rev. Inf. Secur. Priv. **1**(2), 33+ (2011)
34. S. Akbar, K. Nageswara Rao, J.A. Chandulal. Implementing rule based genetic algorithm as a solution for intrusion detection system. Int. J. Comput. Sci. Netw. Secur. **11**(8), 138 (2011)
35. W. Li. Using genetic algorithm for network intrusion detection, in *Proceedings of the United States Department of Energy Cyber Security Group 2004 Training Conference*, Kansas City, Kansas, 2004, pp. 24–27
36. A. Goyal, C. Kumar. GA-NIDS: A Genetic Algorithm based Network Intrusion Detection System. not published. Electrical Engineering and Computer Science, Northwestern University, Evanston, IL (2007)
37. A.C. Divya. GHIDS: a hybrid honeypot system using genetic algorithm. Int. J. Comput. Technol. Appl. **3**, 187 (2012)
38. J.M. Smith. *Evolution and the Theory of Games* Springer US, 1993, pp. 202-215
39. W. Li. Using genetic algorithm for network intrusion detection, in *Proceedings of the United States Department of Energy Cyber Security Group*, 2004
40. S.M. Bridges, R.B. Vaughn. Fuzzy data mining and genetic algorithms applied to intrusion Detection, in *Proceedings of 12th Annual Canadian Information Technology Security Symposium*, 2000, pp. 109–122

Detecting Kernel Control-Flow Modifying Rootkits

Xueyang Wang and Ramesh Karri

1 Introduction

System level virtualization allows users to run multiple operating systems (OS) on a single physical machine. Virtualization can help enterprise users utilize hardware more effectively through server consolidation, and it can make it easy for developers to test programs on different OSes. General computer users can also benefit from this technology [1]. For example, they can concurrently run applications on different OSes.

Although virtualization can improve the system reliability and security by isolating virtual machines (VM) and inspecting VM-based execution, it cannot protect VMs themselves from malicious attacks [2]. A VM may become a victim of viruses and malware just like a traditional non-virtualized machine. Kernel rootkits are formidable threats to computer systems. They are stealthy and can have unrestricted access to system resources. By subverting the OS kernel directly, kernel rootkits are used by attackers to hide their presence, open backdoors, gain root privilege and disable defense mechanisms [3].

Kernel rootkits perform their malicious activities in two ways: modifying the non-control data in the kernel data structures and hijacking the kernel control-flow to conceal resources from system monitoring utilities.

There are not many kernel rootkits of the first category. One example of this kind of rootkit is based on direct kernel object manipulation (DKOM) [4]. The DKOM rootkits hide malicious processes by just unlinking the process objects from the double-linked list for process enumeration. However, by only manipulating kernel objects without modifying the control flow and executing their own code, the function of this kind of rootkit is limited. The DKOM rootkits can only

---

X. Wang (✉) · R. Karri
Polytechnic Institute of New York University, Brooklyn, New York, USA
e-mail: xwang09@students.poly.edu

R. Karri
e-mail: rkarri@poly.edu

R. E. Pino (ed.), *Network Science and Cybersecurity*,
Advances in Information Security 55, DOI: 10.1007/978-1-4614-7597-2_11,
© Springer Science+Business Media New York 2014

manipulate the kernel objects that are in the memory and used for accounting purposes, but is not able to accomplish other rootkits' purposes such as hiding files. Since the modifications only target particular data structures, such as process linked lists, it can be easily located. Techniques for detecting DKOM rootkits have been developed [5, 6].

The kernel rootkits which modify the control-flow are the most common and pose the most threat to system security. A recent analysis [7] indicates that more than 95 % of Linux kernel rootkits persistently violate control-flow integrity. Control-flow modification makes the detection difficult because we do not know what the rootkits will modify. These rootkits may hijack the kernel static control transfers, such as changing the text of kernel functions or modifying the entries of system call table. The control-flow modifying rootkits may also hijack the dynamic control transfers, such as dynamic function pointers.

1.1 Detection Techniques and Limitations

There has been a long line of research on defending against control-flow modi-fying rootkits. Host-based rootkit detection techniques run inside the target they are protecting, and hence are called "in-the-box" techniques. Rkhunter [8] and Kstat [9] detect the malicious kernel control-flow modification by comparing the kernel text or its hash and the contents of critical jump tables to a previously observed clean state. Other techniques try to enforce kernel control-flow integrity by validating the dynamic function pointers that might be invoked during the kernel execution. All the pointers should target valid code according to a pre-generated control-flow graph [10]. Patchfinder [11] applies execution path analysis for kernel rootkit detection. It counts the number of executed instructions by setting the processor to single step mode. In this mode, a debug exception (DB) will be generated by the processor after every execution of the instruction. The numbers counted during the execution of certain kernel functions will be analyzed to determine if the functions are maliciously modified. The main problem with the "in-the-box" techniques is that the detection tools themselves might be tampered with by advanced kernel rootkits, which have high privilege and can access the kernel memory.

With the development of virtualization, the virtual machine monitor (VMM) based "out-of-the-box" detection techniques have been widely studied. VMM-based techniques move the detection facilities out of the target VM and deploy them in the VMM. The isolation provided by virtualization environment signifi-cantly improves the tamper-resistance of the detection facilities because they are not accessible to rootkits inside the guest VMs.

Garfinkel and Rosenblum [12] first introduced virtual machine introspection to detect intrusion. It leverages a virtual machine monitor to isolate the intrusion detection service from the monitored guest. XenAccess [13], VMwatcher [14], and VMWall [15] are virtual machine introspection techniques using memory acquisition. These techniques obtain the guest states from the host side by accessing guest memory pages. However, there is a "semantic gap" between the external and internal observation. To extract meaningful information about the guest state from the low level view of the physical memory state, the detection tools require detailed knowledge of the guest OS implementation. For example, to retrieve the information of a guest VM's process list, the detection tools need to know where this particular data structure is laid out in the guest kernel memory. The location may vary from one implementation to another. Acquiring this detailed knowledge can be a tough task especially when the kernel source code is not available. In regards to security, because the knowledge of the guest OS that the detection tools rely upon is not bound to the observed memory state, those techniques are subject to advanced attacks that directly modify the layout of the guest kernel data structures. The DKSM [16] attack manipulates the kernel structures to fool the detection tools. Moreover, due to the complexity of an OS kernel, there are a large number of kernel objects that might be touched for a certain kernel execution. It is almost impossible to check every single object. When one kernel object is being monitored, an attacker can simply find other targets.

There are also several VMM-based techniques which do not use memory acquisition. NICKLE [17] is a VMM-based approach which defends the kernel rootkits by using a memory shadowing technique. The shadowing memory is isolated by virtualization so it is inaccessible to the guest VM. Only authenticated kernel code in the shadow memory can be executed. NICKLE is effective in preventing unauthorized kernel code from executing in the kernel but cannot protect the kernel control-flow. This makes it susceptible to self-modifying code and "return-into-libc" style attacks. A more important problem is it significantly increases the memory usage. It may double the physical memory usage of a guest VM in the worst case, making it not applicable to virtualization system where many guests are running. Lares [18] monitors a guest VM by placing its hooking component inside the guest OS and protecting it from the hypervisor. These hooks would be triggered whenever certain monitored events were executed by the guest OS. This technique requires modification to the guest OSes, making it not applicable to close-source OSes like Windows.

1.2 HPC-Based "Out-of-the-Box" Execution Path Analysis

To overcome the challenges that the current "out-of-the-box" detection techniques face, we propose an "execution-oriented" VMM-based kernel rootkit detection framework which performs integrity checking at a higher level. It validates the whole execution of a guest kernel function without checking any individual object on the execution path. **Our technique models a kernel function in the guest VM with the number of certain hardware events that occur during the execution. Such hardware events include total instructions, branches, returns, floating point operations, etc.** If the control-flow of a kernel function is maliciously modified, the number of these hardware events will be different from the original one. These monitored hardware events are efficiently and securely counted from the host side by using Hardware Performance Counters (HPCs), which exist in most modern processors.

HPCs are a set of special-purpose registers built into modern microprocessors' performance monitoring unit (PMU) to store the counts of hardware-related activities. HPCs were originally designed for performance debugging of complex software systems. They work along with event selectors which specify the certain hardware events, and the digital logic which increases a counter after a hardware event occurs. Relying on HPC-based profilers, the developers can more easily understand the runtime behavior of a program and tune its performance [19].

HPCs provide access to detailed performance information with much lower overhead than software profilers. Further, no source code modifications are needed. The hardware events that can be counted vary from one model to another. So does the number of available counters. For example, Intel Pentium III has two HPCs and AMD Opteron has four.

HPC-based profiling tools are currently built into almost every popular operating system [20, 21]. Linux *Perf* [22] is a new implementation of performance counter support for Linux. It is based on the Linux kernel subsystem *Perf_event*, which has been built into 2.6+ systems. The user space *Perf* tool interacts with the kernel *Perf_event* by invoking a system call. It provides users a set of commands to analyze performance and trace data. When running in counting modes, *Perf* can collect specified hardware events on a per-process, per-CPU, and system-wide basis.

Our technique takes advantage of both the virtualization technology and the HPC-based profiling. The events are automatically counted by the HPCs, so the checking latency and the performance overhead are significantly reduced. Also, the security is enhanced because the HPCs count the events without a guest's awareness, and they are inaccessible to a guest VM. Since the validation is based on the entire execution flow of a system call, there is no need to check the individual steps on the execution path. It does not matter if the rootkits hijack the static or dynamic control transfers, and it does not matter if the rootkits exploit kernel code injection, function pointer modification, or some other advanced techniques such as self-modifying code and return-oriented attacks. As long as the kernel execution path is persistently changed and the rootkits' own tasks are performed, the number of hardware events that occur during the execution will be different, and can be measured.

2 Design Details

2.1 Threat Model

We target a kernel rootkit which has the highest privilege inside the guest VM. The rootkit has full read and write access to guest VM's memory space, so it can perform arbitrary malicious activities inside the guest VM's kernel space. In order to hide its presence in the guest VM, the kernel rootkit modifies the kernel control-flow and executes its own malicious code. We assume that the VMM is trustworthy. And the rootkit cannot break out of the guest VM and compromise the underlying VMM.

2.2 System Call Analysis Using HPCs

To detect control-flow modifying kernel rootkits, our technique focuses on validating the execution of system calls. System calls are the main interface that a user program uses to interact with the kernel. In order to achieve stealth, a common action that a kernel rootkit performs is to fool the user monitoring utilities (like *ps*, *ls*, *netstat* in Linux). These monitoring utilities retrieve the information about the system states by invoking some system calls. The rootkits usually manipulate the normal execution of these system calls to prevent the monitoring tools from obtaining the correct information. For example, the Linux *ps* command will return the status of all the running processes. The system calls invoked by the *ps* command include *sys_open*, *sys_close*, *sys_read*, *sys_lseek*, *sys_stat64*, *sys_fstat64*, *sys_getdents64*, *sys_old_mmap*, etc. To hide itself and other malicious processes, a rootkit modifies these system calls so that the information about the malicious processes will not appear in the list returned by *ps*. The modifications usually result in a different number of monitored hardware events from the uninfected execution.

To profile the execution of system calls in a guest VM using HPCs, the profiler in the host should have the following capabilities: (1) it should be aware of the occurrence of system calls in a guest VM; (2) it should be able to trigger the HPCs. The existing HPC-based profiling tools cannot meet our design requirements because they are not able to capture the beginning and exit of a system call in a guest VM. So the number of hardware events obtained by a profiling tool cannot be exactly pinned to the execution of a monitored system call. To resolve this issue, our technique connects the profiling tool with the VMM, which is capable of intercepting system calls in the guest VM. The technique can be implemented with any HPC-based profiler. Our proof-of-concept design is based on the Linux *Perf* and the virtualization environment is built with the Kernel-based Virtual Machine (KVM).

Fig. 1 Structure of the
VMM-based rootkit detection
technique with HPCs

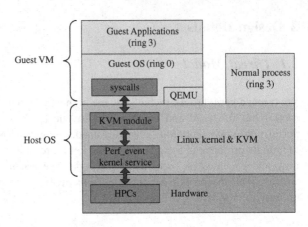

KVM is a full virtualization solution for Linux on hardware containing virtu-alization extensions (Intel VT [23] or AMD-SVM [24] that can run unmodified guest images. In KVM, a guest VM resides in the user space running as a single process and is scheduled like any other normal process. A modified QEMU [25] is used to emulate guest VMs' I/O activities. The processor with hardware virtual-ization extensions has two different modes: host mode and guest mode. The host machine runs in host mode compatible with conventional non-virtualized pro-cessors. The guest VMs run in a de-privileged guest mode. Execution of virtual-ization-sensitive instructions in guest mode will trap to the host, which is called VM-exit. In this way, the host can manage the guests' accesses to virtualized resources. The host maintains a data structure called virtual machine control block (VMCB) to control behaviors of guest VMs. When a guest VM exits to the host, all its CPU states are stored into certain fields located in the VMCB. These states are restored from the VMCB when the CPU switches back from host mode to guest mode.

As shown in Fig. 1, the system calls in a guest VM are intercepted by the KVM module. The KVM module communicates with *Perf_event* kernel service to ini-tialize, enable/disable, read and close HPCs. The checking components, which include the KVM module, the *Perf_event* tool and the HPCs, are deployed outside of the target VM. This isolation prevents the kernel rootkits in the guest VM from compromising the checking procedure. The *Perf_event* kernel service is called to launch a per-process profiling task and enables the HPCs only when a monitored system call is run in the guest VM. By doing so, the counted events are exactly contributed by the execution of the monitored system call in the specific guest VM.

For a given system call, the number of hardware events that occur during the execution varies when different inputs are applied. In our design, to determine if an unusual number of events is caused by the malicious modification to the system call, the counts of hardware events of a monitored system call are compared with those of the corresponding unmodified system call invoked with the same input.

2.3 System Call Interception

To perform the HPC-based check, the execution of a monitored system call should be intercepted by the VMM. System calls are implemented in two ways: interrupt-based system calls and sysenter-based system calls.

Interrupt-based system calls are invoked by executing a software interrupt (INT, with interrupt vector 0x80 for Linux and 0x2e for Windows) instruction, while a kernel can exit from the system call by executing an IRET assembly instruction. The interception of interrupt-based system calls is directly supported by AMD-SVM processors by setting certain bits in the VMCB. Intel VT-extensions currently cannot directly support trapping user interrupts, such as system calls. Ether [26] solves this problem by replacing the system call entry address with an illegal address. The illegal address causes a page fault that can be captured by the VMM. Nitro [27] solves this problem by virtualizing the interrupt descriptor table (IDT).

User space processes enter and exit sysenter-based system calls by executing SYSENTER and SYSEXIT instructions respectively. The sysenter-based system call interception is not directly supported by current hardware assisted virtualization techniques. To implement our design on such platforms, a simple way is disabling CPU features related to sysenter mode in the host OSes to force guest systems to use interrupt-based system calls. Nitro uses another way to achieve this. It captures sysenter-based system calls by injecting system interrupts to guest VMs.

An interrupt vector indicates the type of interrupts (0×00 for divide error, 0×01 for debug, 0×80 for system call, etc.). To determine that a capture of an INT instruction is caused by a system call, the interrupt vector needs to be checked. When a guest VM exits by executing an INT instruction, the address of the instruction is stored in the guest VM's EIP register. By retrieving the address of the INT instruction from the EIP field in the VMCB, we can access guest memory to get the interrupt vector.

Besides capturing the entry and exit of a system call execution, the system call number also should be determined. A system call number is an integer stored in the guest VM's EAX register when a system call is invoked. This value can be obtained from the EAX field in the VMCB.

3 Security Analysis

With the isolation provided by virtualization and the benefits of using HPCs, the execution path analysis is very secure and tamper-resistant. Here, we discuss some possible attacks and show how they can be defended by our technique.

First, the attacks may try to tamper with the counting process. If the event counting is inside the guest VM, the kernel rootkit may disable the counters when its own code is executed and resume the counting when the control-flow returns to the normal execution. In this case, the malicious actions will not be detected since the counts remain the same as the unmodified execution. In our design, the hardware events are counted by the host. The HPCs are out of reach to the rootkits.

Second, the attacks may tamper with the analysis process. Even though the counters are working properly and count all the true numbers, a rootkit may directly manipulate the analysis. Consider Patchfinder, the "in-the-box" execution path analysis technique, as an example. Since the counts are stored in the memory, the kernel rootkits who have full access to the memory can just replace an actual number with a "good" number. For our VMM-based design, the counted events' numbers are read from HPCs by the trusted host and all the analyses are performed by the host. The guest kernel rootkits cannot interfere with the analyses because they do not have access to the host memory.

Although difficult, consider the scenario that an advanced rootkit may modify the execution path but keep the total number of instructions unchanged. Specifically, the rootkit replaces an original function with its own function that executes the same number of instructions. In our HPC-based design, we can monitor multiple hardware events of a guest's execution at the same time. Besides the total number of instructions, we can also count the number of branches, returns, floating point operations, etc. It is extremely hard for a rootkit, almost impossible, to modify a normal execution path with the number of all these events unchanged.

Last, if a clever attacker is aware of the occurrence of a check, it may undo modifications when the check is performed and activate itself again when the check is over. In our design, the detection processes are running in the host without a guest's awareness. The only thing the guest can see is the execution of a test program. However, from the guest's point of view, the execution of a test program is no different from the execution of other programs. So a guest is not able to know when it is being monitored. Additionally, we can randomize the intervals between checks to avoid attackers' prediction of checking period.

The only limitation of our technique is detecting short-life kernel modifications. Nevertheless, the reason why attackers employ a rootkit is that they try to perform long-term functionality in the target system without being discovered. So they need a rootkit to provide long-term stealth for the malicious activities. According to this basic goal, a short-life rootkit is useless. Since our technique focuses on detecting kernel rootkits not other short-life attacks, this limitation can be neglected.

4 Evaluation

To evaluate the effectiveness of our technique, we test our technique with a real-world kernel rootkit: SUCKIT 1.3b [28].

SUCKIT, Super User Control Kit, is one of the widely known kernel rootkits. It runs under Linux and is used to hide processes, hide files, hide connections, get

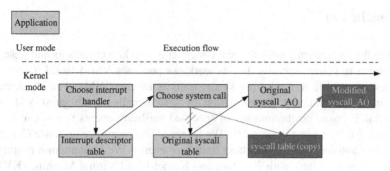

Fig. 2 Execution flow of the SUCKIT kernel rootkit

Table 1 Number of hardware events counted from the execution of original and infected system calls

Events monitored		System calls monitored		
		sys_open	sys_read	sys_getdents64
Retired instructions	Original	1081	370	1808
	Infected	10119	590	6200
Retired returns	Original	17	6	24
	Infected	132	15	140
Retired branches	Original	191	78	160
	Infected	2663	104	1805

root privilege, etc. It replaces the system call table with its own copy, and then uses its own system call table to redirect to the malicious system calls. Figure 2 shows the execution flow of the SUCKIT kernel rootkit.

In our evaluation, SUCKIT 1.3b is run on a Redhat 7.3 guest VM with Linux kernel 2.4.18 and the host runs Ubuntu 11.10 with Linux kernel 3.0.16. We check three system calls, *sys_open, sys_read, sys_getdents64*, which are usually the targets for a kernel rootkit. Three hardware events, retired instructions, retired returns, and retired branches, are monitored simultaneously for the execution of each system call. The counted number of an infected system call is compared with the number of the original one. Table 1 shows the experimental results.

From the results, we can see that in order to introduce their own functionality, the rootkits usually significantly modify the original system calls. The difference in the number of events between normal and infected executions is very notable. For SUCKIT 1.3b, all the three monitored system calls are maliciously modified. The system call *sys_open* is modified most heavily. The number of hardware events counted from the infected *sys_open* is more than ten times as the original one.

5 Conclusion

Control-flow modifying rootkits are the most common kernel rootkits and pose the most threat to system security. In this work, we present a VMM-based framework to detect control-flow modifying kernel rootkits in guest VMs. The checking is performed by validating the execution of system calls in the guest VM. The validation is based on the number of specified hardware events that occur during the execution of a guest system call. We leverage Hardware Performance Counters (HPCs) to automatically count these hardware events. We implement a prototype of our design on Linux with *Perf* tool and Kernel-based Virtual Machine (KVM). Our experimental results demonstrate its practicality and effectiveness.

Acknowledgments This material is based upon work funded by AFRL under contract No. FA8750-10-2-0101. Any opinions, findings and conclusions or recommendations expressed in this material are those of the author(s) and do not necessarily reflect the views of AFRL.

References

1. K. Nance, M. Bishop, B. Hay. Virtual machine introspection: observation or interference? in *IEEE Symposium on Security and Privacy*, September 2008, pp. 32–37
2. F. Azmandian, M. Moffie, M. Alshawabkeh, J.G. Dy, J.A. Aslam, D.R. Kaeli, Virtual machine monitor-based lightweight intrusion detection. ACM SIGOPS Oper. Syst. Rev. **45**(2), 38–53 (July 2011)
3. Z. Wang, X. Jiang, W. Cui, P. Ning, Countering kernel rootkits with lightweight hook protection, in *Proceedings of the 16th ACM Conference on Computer and Communications Security*, November 2009, pp. 545–554
4. G. Hoglund, J. Butler, *Rootkits: Subverting the Windows Kernel* (Addison Wesley, Boston)
5. G. Hunt, D. Brubacher, Detours: Binary interception of win32 functions, in *Proceedings of the 3rd USENIX Windows NT Symposium*, 1999, pp. 135–143
6. J. Rutkowska, Klister, http://www.rootkit.com/vault/joanna/klister-0.4.zip
7. J.N.L. Petroni, M. Hicks, Automated detection of persistent kernel control-flow attacks, in *Proceedings of ACM Conference on Computer and Communications Security*, 2007, pp. 103–115
8. Rkhunter, http://packetstormsecurity.org/files/44153/rkhunter-1.2.8.tar.gz.html
9. Kstat-kernel security therapy anti-trolls. http://www.s0ftpj.org/en/tools.html
10. M. Abadi, M. Budiu, U. Erlingsson, J. Ligatti, Control-flow integrity, in *Proceedings of ACM Conference on Computer and Communications Security*, November 2005
11. J. Rutkowska, Execution path analysis: finding kernel based rootkits. Phrack Mag. Vol. 0x0b, NO. 0x3b, Phile #0x0a, 2003
12. T. Garfinkel, M. Rosenblum, A virtual machine introspection based architecture for intrusion detection, in *Proceedings of Network and Distributed Systems Security Symposium*, 2003, pp. 191–206
13. B. Payne, M. de Carbone, W. Lee, Secure and flexible monitoring of virtual machines, in *Computer Security Applications Conference*, December 2007, pp. 385–397
14. X. Jiang, X. Wang, D. Xu, Stealthy malware detection through vmm-based out-of-the-box semantic view reconstruction, in *Proceedings of Acm Conference on Computer and Communications Security*, November 2007, pp. 128–138

15. A. Srivastava, J. Giffin, Tamper-resistant, application-aware blocking of malicious network connections, in *Proceedings of International Symposium on Recent Advances in Intrusion Detection*, 2008, pp. 39–58

16. S. Bahram, X. Jiang, Z. Wang, M. Grace, J. Li, D. Xu, Dksm: Subverting virtual machine introspection for fun and profit, in *Proceedings of the 29th IEEE International Symposium on Reliable Distributed Systems*, October 2010

17. R. Riley, X. Jiang, D. Xu, Guest-transparent prevention of kernel rootkits with vmm-based memory shadowing, in *Proceedings of International Symposium on Recent Advances in Intrusion Detection*, 2008

18. B. Payne, M.S.M. Carbone, W. Lee, Lares: an architecture for secure active monitoring using virtualization, in *Proceedings of the IEEE Symposium on Security and Privacy*, May 2008

19. J. Du, N. Sehrawat, W. Zwaenepoel, Performance profiling of virtual machines, in *Proceedings of the 7th ACM SIGPLAN/SIGOPS International Conference on Virtual Execution Environments*, July 2011

20. I. Inc. Intel vtune performance analyzer, http://software.intel.con/en-us/intel-vtune, 2010

21. J. Levon, P. Elie, Oprofile: a system profiler for linux, http://oprofile.sourceforge.net, 2010

22. Performance counters for linux, http://lwn.net/Articles/310176, 2010

23. Intel virtualization technology: hardware support for efficient processor virtualization. August 2006

24. Amd secure virtual machine architecture reference manual, May 2005

25. www.qemu.org/

26. A. Dinaburg, P. Royal, M. Sharif, W. Lee, Ether: malware analysis via hardware virtualization extensions, in *Proceedings of ACM conference on Computer and Communications Security*, October 2008

27. J. Pfoh, C. Schneider, C. Eckert, Nitro: hardware-based system call tracing for virtual machines. Adv Inform Comput Secur **7038**, 96–112 (2011)

28. Sd, Devik, Linux on-the-fly kernel patching without LKM. Phrack Mag **11**, 2004

Formation of Artificial and Natural Intelligence in Big Data Environment

Simon Berkovich

1 Introduction: Big Data Algorithmics in the Laws of Nature

The amount of information associated with Life and Mind overwhelms the diversification of the material world. Disregarding the information processing in the foundation of Nature modern science gets into a variety of complications. The paradigm of fundamental physics that does not explicitly incorporate an information processing mechanism is not just incomplete, it is merely wrong. As John A. Wheeler tersely said: "the physical world is made of information with energy and matter as incidentals".

Realization of information processing encounters two types of problems related to hardware and to software. In this paper, we contemplate in a broad sense the software problems in connection to the situation of the inundation of information dubbed Big Data. The hardware problems associated with this Big Data situation have been addressed in some general way in our previous publications. As an issue of practical computer engineering these problems has been outlined in [1]. The hardware model of the informational infrastructure of the physical world in the form of a cellular automaton mechanism has been described with numerous ramifications in [2–5].

Constructive solutions for natural science necessitate elegant operational algorithms. Any algorithm is workable, but inappropriate algorithms translate into clumsy ad hoc theories. The ingenious algorithmic solutions devised for the Big Data situation transpire as effective laws of nature.

The Big Data situation stumble upon two types of problems: how to exercise meaningful actions under overabundance of information and how to actually generate objects having extremely rich information contents. Starting with Sect. 2,

S. Berkovich (✉)
Department of Computer Science, The George Washington University,
Washington, DC 20052, USA
e-mail: berkov@gwu.edu

R. E. Pino (ed.), *Network Science and Cybersecurity*,
Advances in Information Security 55, DOI: 10.1007/978-1-4614-7597-2_12,
© Springer Science+Business Media New York 2014

we introduce a computational model for Big Data that goes beyond ordinary Turing computations. The incapability to explicitly use all the available Big Data leads to the concept of bounded rationality for Artificial Intelligence as depicted in Sect. 3. This approach emphasizes the Freudian idea of the decisive role of unconsciousness for Natural Intelligence, which is delineated in Sect. 4.

Generating bulkiness through step-by-step growth is not suitable for mass production. Thus, creation of Mind implicates usage of Cloud Computing as described in the above-mentioned Sect. 4. In this case, the tremendous contents of human memory are built-up through joining an already existing repository of information. Section 5 considers both types of Big Data developments. Mental problems, like neuroses, schizophrenia, and autism, are believed to present pure software distortions of the context background brought about by Cloud Computing. Also considered is the other part of massive Big Data formations, which is the method for self-reproduction of macromolecule configurations.

The suggested Big Data algorithms can be fulfilled in the physical world organized as an Internet of Things. Section 6 concludes with an overview of experimental possibilities to verify this surmised construction. It presents the most compelling *Experimentum Crucis* exposing the Internet of Things in the framework of the Holographic Universe.

2 The Computational Model for Big Data

Information processing begins with the idea of a computational model. Computational model is an abstract scheme for transforming information. It operates in the following cycling: extracting an item of data from the memory—transforming the given item of data—returning the transformed item to the memory. The first idea of this kind with memory presented as an infinite tape having sequential access was introduced by Alan Turing as a formal definition of the concept of an algorithm. John von Neumann had introduced a practical computational model using random access memory for the realization of first computers. Remarkably, despite of the tremendous successes of computer technology at all fronts in more than half a century the basic computational model stays the same. This indicates at something of fundamental significance. The famous Church-Turing Thesis conveys an informal statement that all reasonable computational models are in fact equivalent in their algorithmic expressiveness. In simple words, any calculation that can be done on one computer can be done on another computer; the difference is only in performance. This immediately raises the question about the brain. Thus, on one hand the facilities of the brain have to be equated with those of a conventional computer, on the other hand, this does not seem likely.

We introduce a somewhat different computational model specifically suitable for a Big Data environment (Fig. 1). In this computational model the extraction of a data item from memory is determined by the whole bulk of data. Thus, only a relatively small part of the given Big Data explicitly contributes to actual

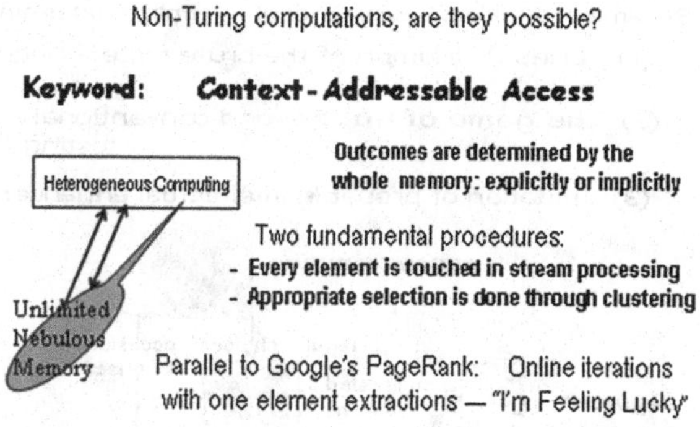

Fig. 1 The "Big Data"
computational model

Non-Turing computations, are they possible?

Keyword: Context-Addressable Access

Heterogeneous Computing

Unlimited
Nebulous
Memory

Outcomes are determined by the
whole memory: explicitly or implicitly

Two fundamental procedures:
- Every element is touched in stream processing
- Appropriate selection is done through clustering

Parallel to Google's PageRank: Online iterations
with one element extractions — "I'm Feeling Lucky"

computations, the vast majority of these Big Data contributes to the computations implicitly by determining what data items should be extracted for definite usage. So, access to specific data items in this computational model is determined by the context of all data. This context-addressable access is different from that in traditional associative memories. It is similar to what is provided by Google's PageRank algorithm. Some hardware/software details for the realization of the presented computational model are discussed in [1].

3 Bounded Rationality Approach to Artificial Intelligence

Potentials of Turing computations may be expanded with a speculative assumption of an "Oracle"—a black box guiding the choice of available alternatives. In complexity theory, an "Oracle machine" is an abstraction to study decision problems. The lofty question of $P = NP$, i.e., whether non-deterministic and deterministic decision problems are equivalent in their efficiency is not strictly resolved yet, but the extraordinary power of "Oracle" computations is quite obvious.

The Big Data computational model exhibits the features of an "Oracle" machine". The selections of appropriate data items by a genuine "Oracle" or by a huge context—"pseudo-Oracle" are indistinguishable. The alternative: a truly supernatural "Oracle" vs. simulated "pseudo-Oracle" can be compared with the alternative: "Free Will" versus "Determinism". Thus, random choices with really random generators or with pseudo-random procedures, as well as life or pre-recorded TV shows, are indistinguishable.

The problem of "Artificial Intelligence" is usually associated with making clever decisions under an abundance of data, real or synthetic. In general, this might involve creating a very elaborate model for the system of study, so it would be able to accommodate as much as possible of all of the available data. However, in many practical cases this is unrealistic. A more sensible approach would be to utilize a simplified model of the system, which is guided by an "Oracle". An exemplar of an "Oracle" could be produced within the suggested computational

Scheme of thinking: simple model --- context-determined selection of alternatives

(1)   **Chess**: triumph of the brute force - Deep Blue vs. Garry Kasparov

(2) **The game of Go**: beyond conventional computer facilities —
                                            testing the "Big Data" approach

(3) Imitation of probable intellectual enhancement

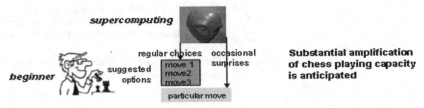

Fig. 2 Decision-to-data problem

model using a rich context of Big Data. So, this "Artificial Intelligence" system could acquire "Intuition from Context" (Fig. 2).

The classical target for Artificial Intelligence is the game of chess. The success in this direction has been achieved primarily by application of high computational power. With the suggested approach we are planning to test another scheme: a beginner displays several possible moves in accordance with some simplified understanding of the game; an "Oracle" (a qualified human or a supercomputer) makes a best selection of the displayed moves. Thus, a substantial playing improvement could be anticipated. It would be interesting to explore this approach for the game of Go that is more computationally challenging than the game of chess. A possible beneficial influence of this arrangement on the mental state of the implicated human player will be brought up in Sect. 5.1.

4 Realization of Natural Intelligence

In 1943 McCulloch and Pitts introduced a formal computational unit—an artificial neuron. An elaborate network of these units is able to solve intricate multidimensional mathematical problems [6]. At the same time, there is a strong belief that complex artificial neural network activities should exhibit the sophistication of the brain. Yet the approach to the conception of the brain as a "complex" network of neurons is inadequate, since slow erratic combinations of electrical and chemical processes in neuron systems cannot match the high performance characteristics of the brain in terms of processing power and reliability. It becomes apparent that understanding of the brain needs a radical paradigm shift towards extracorporeal organization of human memory [7]; also, see the analysis in [8]. Extracorporeal realization of biological memory is based on our cellular automaton model of the physical world resulting in the organization of Nature as an Internet of Things [2–5, 9]. Corresponding illustrations are given below in Figs. 3

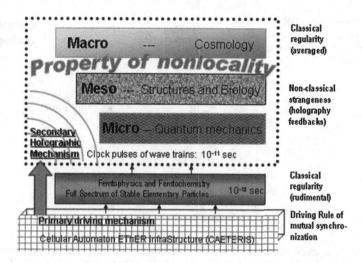

Fig. 3 Cellular automaton unfolding into the holographic universe

and 4. Since the organization of the brain operates with tremendous amounts of information its workings should be presented within a construction that is able to handle the suggested computational model for Big Data.

The information capacity of human memory should be virtually unlimited as everything is continuously recorded and never erased. From this perspective John von Neumann estimated that the capacity of human memory is about 2.8×10^{20} bits [10]. The tremendous amount of information stored in human memory is used implicitly as a passive context, while only a rather small portion of this information is active explicitly. In the book [11], it is somehow estimated that humans of 80 years age actively employ only a very tiny piece of all memory information—about 1 Gb. Influences of the passive background on the workings of the brain are in accordance with Freud's theory of unconsciousness. The role of the unconsciousness in mental diseases is discussed in the next Sect. 5.1.

The computational model with context-addressable access could be beneficial for a broad number of applications when the information processing performance increases with the accumulation of examples. Particular instances include learning a language, pattern recognition, reinforcing the skills etc. Building up a large context allows approaching the effective solutions of almost all problems. However, some of the information processing tasks, for example, arithmetic calculations, would be done more smoothly with ordinary computational models rather than employing Big Data contexts.

The main operational mechanism for the implementation of the Big Data computational model is streaming. The significance of the streaming capacity for the organization of the brain is exposed through the effect of the so-called Penfield movies [12]. This effect was observed by stimulation of different parts of the brain during surgery. The subject of this stimulation began to relive earlier periods of time in the greatest detail including various sensory components—visual, auditory,

Fig. 4 Layered holographic
memory: origin of
nonlocality and mesoworld
sophistication

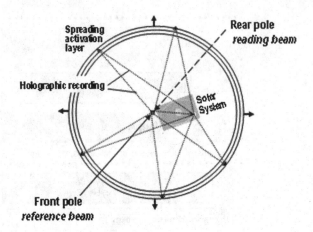

olfactory, and even somatic. Two circumstances are relevant to our Big Data consideration: first, the recall produces random samples of true experience, usually, of no significance in the life of the patient—context background, and second, the appearing pictures are "time-ordered", the events go forward but never backwards—this enables the organization of streaming.

The principles of holography materialize the extracorporeal placement of human memory. Holographic organization of the Brain and the Universe is a popular topic for abstract theoretical speculations (see [13]). In our concept, the holographic mechanism is a secondary construction atop of the cellular automaton model of the physical world (Fig. 3). Realization of a holographic mechanism entails clear technical requirements: a reference beam generating wave trains pulses and a relatively thin recording medium in compliance with spatial and temporal coherence. This leads to a special design of holographic memory with spreading recording layer (Fig. 4). The presented construction naturally incorporates the otherwise inconceivable property of the Universe nonlocality.

The spreading activation layer of the holographic memory acquires and retains signals from all the events in the Universe. Among those are signals from the brains that are recorded as the states of its memory. This information is modulated by the conformational oscillations of the particular DNAs, so the whole holographic memory of the Universe is shared among the tremendous variety of biological organisms [14].

We would like to single out two prominent physical properties in relation to the considered construction: the tridimensionality of space and the anisotropy of the Cosmic Microwave Background. As long as physical and biological processes rely on the informational control of the holographic mechanism it is necessary that the waves involved in this mechanism propagate in accordance with the Huygens principle, i.e., with sharply localized fronts. Otherwise, the interference of holographic waves will blur. Huygens principle occurs strictly only in 3D space; this implicates the tridimensionality for the physical space and, hence, for the space of perception [15].

The appearance of the anisotropy of the Cosmic Microwave Background is of remarkable significance. To a very great surprise of cosmologists, in about the year of 2003 a certain pattern built-in in the Cosmic Microwave Background has been discovered [16]. This pattern was called the Axis-of-Evil as it merely should not be there—as commonly understood the Cosmic Microwave Background must be uniform. There had been put forward a number of esoteric ideas that the unexpected imprint in the Cosmic Microwave Background is a message from a Supreme Being or from a neighboring Universe. In our theory, the Cosmic Microwave Background is not a post-creation remnant of the cooling down matter, but an accompanying factor of the layered holographic activities. Our explanation of the anisotropy of the Cosmic Microwave Background Cosmic is natural, easy, and neat. The Cosmic Microwave Background is indeed uniform if observed from the center—the pole issuing reference beam. But when observed from the eccentric position of the Solar system these activities become distorted. Our model exactly predicts the angle between the dipole and quadrupole axes: −40° [17] (see Fig. 5). If necessary, higher order axes can be also exactly calculated and compared. Another, more simple and clear manifestation of the Holographic Universe is referred to in the conclusion.

Full realization of the surmised computational scheme for the organization of the brain with the required holographic memory parameters does not seem realistic with the hardware resources available on Earth. As long as the major operation needed for the organization of the brain is massive stream processing, partial realization of this functionality for the suggested Big Data computational model can go in two directions.

First, following the way suggested in [1] the required stream processing could be arranged with the pipelining that has a distinctive capability to effectively accommodate on-the-fly computations for an arbitrary algorithm. The most essential part of this processing is the suggested technique for on-the-fly clusterization. This type of the brain functionality would be most suitable for special intelligent tasks, such as knowledge discovery—formulation and verification of hypotheses.

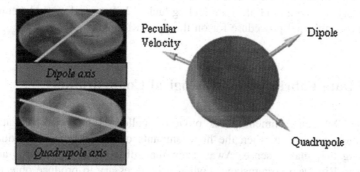

Fig. 5 Anisotrophy of the cosmic microwave background "Axis-of-Evil"—eccentric observation of holographic recordings predicated theoretically several years before the actual discovery

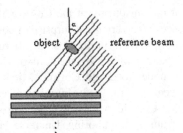

* Hardware resources are not available → Software simulation --- Cloud Computing
* Employing microprocessor pipeline for stream computations --- On-the-fly clustering
* An extra operational facility --- Random address access using angle α as an address
* Sophisticated partial matches for "Knowledge Discovery" --- Exchanging background

Fig. 6 Emulation of cognitive facilities of the brain with digital holography

Second, in a much broader sense, the imitation of Natural Intelligence could be achieved by direct implementation of the basic holographic scheme for the brain by emulation of the smart unobtainable hardware of the Universe with digital holography. This can be approached with Cloud Computing (Fig. 6). At a given Internet site "layers" of the holographic transformations for different objects are calculated with different angles of the incident reference beam. Search for a specified object is done by a sequential lookup for best matches with the digital holograms in the recorded layers. The incidence angle of the reference beam reconstructed even from partial match identifies the object. The computational process consists of iterations of these sequential lookups.

The mental activities of the brain are supposed to be completely software-programmable with such a Cloud Computing arrangement. The characteristic feature of "subconscious" processing—manipulation with small quantities of data whose selection is holistically determined by the entire data contents—can be exactly accommodated in the given framework. The selection procedure can be paralleled with an iterative version of Google's PageRank, where a uniquely specified item rather than a subset of items must be extracted. This specification may simply rely on a kind of "I am feeling lucky" button, and might be enhanced using the established procedure for on-the-fly clustering.

5 Big Data Fabrication of Biological Configurations

The Big Data environment poses obvious challenges to the organization of information processing when the huge amounts of data have to be reduced to something that makes sense. Away from that, different complications arise in relation to Big Data circumstances when it is necessary to produce objects with tremendous structural variety. This is an important existential question. A viable object has to be created in a relatively small number of steps, say in O(1) using algorithms terminology. Algorithmic constructions that take O(N) steps for creating very large objects would not be practicable. A productive resolution of the Big Data induced concerns presents a decisive issue for bio-medical objects. Here we will show two characteristic instances of these Big Data problems that can be effectively resolved within our concept of the physical world as an Internet of Things, as it combines informational and physical processes. These two Big Data creations are related to the informational filling of human memory and to the reproduction of the material variety of macromolecules.

5.1 Joining the Cloud and Mental Disorders

Let us consider how human memory could be amassed with the Big Data. Human brain contains about 10^{11} neurons and 10^{14} synapses. It is believed that updates of the synapses somehow develop the contents of human memory. Let us assume that chemical processes associated with one update take 1/100 s. So, getting one update at a time would lead to formation of the whole system of synapses in about 30,000 years. Thus, assuming that a child of 3 years of age acquires a system of synapses, which is in essence prepared, this system should have been continuously reorganized with the pace of 10^6 updates per second.

In terms of algorithmic effectiveness, formation of a Big Data memory structure by individual updates, even performed in parallel, does not appear feasible. In our conception of the physical world as an Internet of Things the problem of the formation of biological memory is efficiently resolved in a simple way with much less time and effort. This can be achieved merely by joining the holographic Cloud. The required updates of the Cloud contents are done at the pace of the repetition rate of the holography reference beam—10^{11} Hz. The information substance obtained from the Cloud basically constitutes the background context for the Big Data computational model. Evolution of this context while transferring from one generation to the next one is a conservative process. For good or for worse, the core of this context—paradigms, habits, myths, morality, etc.—cannot undergo rapid transformations. This context changes slowly. In a sense, the fact of contest conservatism keeps up with the von Neumann's saying: "It is just foolish to complain that people are selfish and treacherous as it is to complain that the magnetic field does not increase unless the electric field has a curl. Both are laws of nature".

Possible disruptions of the considered process for acquiring the context background for newly developed organism can result in mental disorders, like neuroses, schizophrenia, and autism. In most cases of such disorders changes in the physical constituents of the brain are insignificant. So, it is a software rather than a hardware problem.

Various details associated with the considered mental disorders seem to corroborate our hypothetical scheme of their origin. First, let us start with the issue of heredity. The article [18] reports a sensational observation that "older men are more likely than young ones to father a child who develops autism or schizophrenia". The study found that "the age of mothers had no bearing on the risk for these disorders". The explanation of this observation implicates "random mutations that become more numerous with advancing paternal age". It is questionable that the alleged mutations occur at random because of the doubts why should these mutations target specifically mental disorders. In our concept, the observed effect can be elucidated considering the diagram in Fig. 4: the amount of the holographic layers accumulating the father's life information increases with father's age; so when this information is used to create the context background for the newborn child it might encounter more disruption influences. Also, the suspected transgeneration epigenetic influences on autism could be related to the same surmised mechanism for the context background formation. Very surprisingly, as indicated in [19], "The mental health of a child's mother during pregnancy is widely considered a risk factor for emotional and behavioral problems later in the child's life. Now a new study finds that the father's mental health during the pregnancy also plays a role."

More than 500 genes have so far been implicated in autism showing that no clear genetic cause will be identified [20]. Thus, it is vital to look at the role of the environmental factors. Babies exposed to lots of traffic-related air pollution in the womb and during their first year of life are more likely to develop autism, according to [21]. In our view, nanodust affecting DNA conformational oscillations, and, hence, their communicating facilities, changes the context background. Finally, let us turn our attention to some possibilities of recovery as reported in [22]: "Doctors have long believed that disabling autistic disorders last a lifetime, but a new study has found that some children who exhibit signature symptoms of the disorder recover completely." In our concept, this self-cure could be enhanced by applying the technique exhibited in Fig. 2.

5.2 3D Printing and Self-Replication of Macromolecules

The organization of the physical world as an Internet of Things allows Big Data configurations to be produced not just for informational structures but for material constructions as well. The former are being developed through joining the Cloud Computing process, while the latter making use of quantum mechanics provide what can be called quantum "3D printing". Thus, an impact of information signals on material activities is exercised in synapses gaps where the propagation of electro-chemical pulses in axons and dendrites continues by chemical neurotransmitters. This way neural activity inside the brain can be modulated by the information control from the outside extracorporeal memory.

A vital Big Data operation in living systems is self-replication of macromolecules. This is largely related to the creation of proteins in morphogenesis and metabolism. The regular way of protein production according to the Central Dogma of molecular biology: DNA—mRNA—protein is not sufficient. Two main reasons can be pointed out. First, it is not feasible to have bulkiness fabricated step-by-step.

And second, in many circumstances the proteins are to be exactly reproduced with their folding structures, like prions in the case of "Mad Cow disease". The other way for macromolecules reproduction that we will present here has been suggested in [23]. The Big Data malfunctions associated with protein reproductions constitute for the brain "hardware" problems—neurological diseases, while above mentioned disruptions associated with the creation of contextual background constitute for the brain "software" problems—mental disorders.

The suggested procedure of self-replications of macromolecules is depicted in Fig. 7. It is based on our interpretation of quantum mechanics behavior as a result of interactive holography [24]. The involvement of the holographic mechanism directly exposes the dominant quantum property of nonlocality that otherwise appears inconceivable. The specifics of the quantum mechanics behavior are essentially determined by the interaction of two entities: the actual particles and their holographic feedback images. It has been shown that quantum transitions as random walks of these entities are described by Schrödinger's equation. The imprecision in localization of a particle between actual and virtual entities leads to the fundamental quantum principle of uncertainty. In relation to macromolecules this produces mesoscopic displacements of their components that leads to an effective algorithmic procedure for reproduction of the "Big Data" structures.

The facilities for self-reproduction possibility of macromolecules should reveal new yet not recognized properties of the physical world as anticipated by P.L. Kapitsa [25]. The surmised algorithm for self-replication of macromolecules develops by means of swapping of particles with their holographic placeholders as illustrated and explicated in Fig. 7. The suggested self-replication algorithm can be figuratively imagined as "Xerox" copying. The proliferation of proteins in bio-logical organisms by means of application of this algorithm is analogous to the creation of Gutenberg's Galaxy of books thanks to a breakthrough invention of the printing press.

Fig. 7 The algorithm for reproduction of macromolecules. *1* Macromolecule components with holography copies. *2* Random scattering of the components over both place. *3* Two half-full patterns are reconstructed to completeness

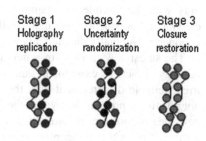

Stage 1
Holography
replication

Stage 2
Uncertainty
randomization

Stage 3
Closure
restoration

6 Conclusion: *Experimentum Crucis*

Coping with the "Big Data" situation constitutes the key problem for purposeful behavior in natural and artificial systems. Human reaction to the Big Data environment is bounded rationality—a decision-making process complying with cognition limitations and imposed deadlines. The ideology of bounded rationality leads to a computational model of the brain that goes beyond the traditional Turing algorithmics. This Big Data computational model reveals the unconsciousness as the basis for sophistication.

The effectiveness of the given computational model encourages evaluating this approach for the general paradigm of the organization of biological information processing. Such a consideration leads to the view of the physical world as an Internet of Things. This kind of theoretical edifice is inspired by the practical advancements in modern information technology. Similarly, creation of the steam engine in the Industrial Revolution promoted the theory of thermodynamics. The new paradigm of the physical world as an Internet of Things materializes in the framework of the Holographic Universe. Distinctively, information processes in this construction realize the most mysterious property of the physical world—quantum nonlocality. Nowadays, conventional interpretation of quantum theory encounters more and more serious complications. Thus, several prominent scientists say that "the absurdity of the situation" cannot be ignored any longer and quantum mechanics is going to be replaced with "something else" [26].

Introduction of a new idea encounters fierce opposition from the public in general, and this work should not be an exclusion. Yet—an exceptional circumstance—this work clearly shows why opposition to new ideas actually happens. People do not debate about the validity of arguments as logic is integrated in human mental process. People argue about the interpretations of premises that are determined by the scheme of the built-up "Big Data" context background. Therefore, there is basically no chance to make people to change their mind. The famous words of Max Plank manifestly present the pessimistic reality in conformity with the considered scheme: "A new scientific truth does not triumph by convincing opponents and making them see the light, but rather because its opponents eventually die, and a new generation grows up that is familiar with it."

The surest way to confirm a new theory is *Experimentum Crucis*. This methodology attracts an experiment that is consistent with the new theory, but is in an irreconcilable disagreement with the established theory. Overall, experiments do not positively prove a theory; but they can only surely disprove it. So, as long as a new idea cannot prevail directly, it can do it by a counterattack with an *Experimentum Crucis* that undermines opponent's paradigm.

As a matter of fact, holography mechanism is quite sensitive to objects locations. Thus, the eccentric positioning of the Solar System in the Holographic Universe (Fig. 4) determines the otherwise incomprehensible anisotropy of the Cosmic Microwave Background. Yet, a more compelling *Experimentum Crucis* for the establishment of the given construction should be simple and sensible. As

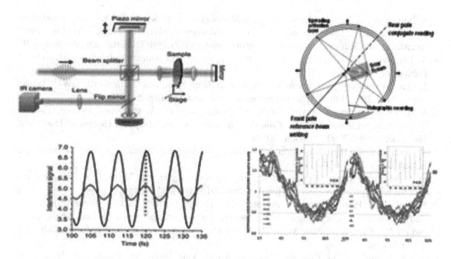

Fig. 8 Revealing the holographic infrastructure of the universe: parallels of the Michelson experiment and the calendar effect

such, we consider the "calendar effect" introduced in [23]. As seen in Fig. 4, the position of the Earth changes due to its motion on the solar orbit; so we can expect annual variations in all phenomena that are related to quantum mechanics. This "calendar effect" is of universal applicability, and it is apparently clear, like, for example, the statement that nearly all bodies expand when heated. Currently, most vivid examples of the surmised calendar effect have been determined for two phenomena: annual variability of the rates of radioactive decay in physics [27] and "seasonal" variations in cardiac death rates in biology [28]. Less clear-cut examples of calendar effect for numerous bio-medical occasions have been described; as to physics, these calendar variations have to be anticipated for a number of fine quantum effects whose outcomes should systematically fluctuate from month to month.

The most articulate manifestation of the calendar effect through variations of heart attacks [28] can be regarded as a generalization of the celebrated Michelson experiment, at this time with a positive outcome, where holography plays the role of interferometry and ailing hearts serve as detectors for malformed proteins (Fig. 8).

References

1. S. Berkovich, D. Liao, On clusterization of "big data" streams, in *COM.Geo '12 Proceedings of the 3rd International Conference on Computing for Geospatial Research and Applications* (ACM, New York, 2012)
2. S. Berkovich, in *Mutual Synchronization in a Network of Digital Clocks as the Key Cellular Automaton Mechanism of Nature. Computational Model of Fundamental Physics* (Synopsis, Rockville, 1986)

3. S. Berkovich, in *Cellular Automaton as a Model of Reality: Search for New Representations of Physical and Informational Processes*, (Moscow University Press, Moscow, 1993) (translated into Russian), http://www.chronos.msu.ru/RREPORTS/berkovich_avtomaty.pdf

4. S. Berkovich, H. Al Shargi, *Constructive Approach to Fundamental Science* (University Publishers, San Diego, 2010)

5. J. Qin, Elementary particles of matter in a cellular automaton framework. GWU, Master thesis, May 2012, http://pqdtopen.proquest.com/#abstract?dispub=1510409

6. V.M. Krasnopolsky, *The Application of Neural Networks in the Earth System Sciences. Neural Network Emulations for Complex Multidimensional Mappings* (Springer, Dordrecht, Heidelberg, New York, London, 2013)

7. S. Berkovich, On the information processing capabilities of the brain: shifting the paradigm. Nanobiology **2**, 99–107 (1993)

8. D.R. Forsdyke, Samuel Butler and human long term memory. J. Theor. Biol. **258**, 156–164 (2009), http://post.queensu.ca/ ~ forsdyke/mind01.htm

9. S. Berkovich, Physical world as an Internet of Things, in *COM.Geo '11 Proceedings of the 2nd International Conference on Computing for Geospatial Research and Applications*, Article No. 66, (ACM, New York, 2011), www.ogcnetwork.net/system/files/Berkovich_220-397-1-RV.pdf

10. J. von Neumann, *The Computer and the Brain*, Yale University Press: New Haven and London, 2000

11. E. Libbert (ed.), *Kompenduium Der Allgemeinen Biologie* (VEB Gustav Fisher Verlag, Jena, 1982)

12. W. Penfield, *The Mystery of Mind* (Princeton University Press, Princeton, 1975)

13. M. Talbot, *The Holographic Universe* (Harper Perennial, New York, 1991)

14. S.Y. Berkovich, *On the "Barcode" Functionality of DNA, or the Phenomenon of Life in the Physical Universe*, (Dorrance Publishing Co, Inc., Pittsburgh, 2003), (a shorter version has been presented at http://arxiv.org/abs/physics/0111093)

15. S.Y. Berkovich, The dimensionality of the informational structures in the space of perception (posing of problem). Biophysics **21**, 1136–1140 (1976), http://structurevisualspacegroup. blogspot.com/2010/10/holographic-model-of-human-memory-and.html

16. M. Chown, 'Axis of evil' warps cosmic background. New Sci. 22 October 2005(2552) p.19ff

17. S.Y. Berkovich, J. Favre, An eccentric view on the kinematical scheme of the big bang and the absolute impression of the cosmic background radiation. A paper in [3], pp. 235–287

18. B. Carey, Father's age is linked to risk of autism and schizophrenia, http://www.nytimes.com/2012/08/23/health/fathers-age-is-linked-to-risk-of-autism-and-schizophrenia.html?_r=0

19. M. Healy, Expectant dads' mental health linked to kids' behavior, http://www.usatoday.com/story/news/nation/2013/01/04/dads-mental-health-prenatal/1807709/

20. R. Bazell, Experts: wide 'autism spectrum' may explain diagnosis surge, http://vitals.nbcnews.com/_news/2012/03/29/10925941-experts-wide-autism-spectrum-may-explain-diagnosis-surge?lite

21. A.M. Seaman, Traffic pollution tied to autism risk: study, http://www.reuters.com/article/2012/11/26/us-traffic-pollution-autism-idUSBRE8AP16020121126

22. B. Carey, Some with autism diagnosis can overcome symptoms, study finds, http://www.nytimes.com/2013/01/17/health/some-with-autism-diagnosis-can-recover-study-finds.html

23. S. Berkovich, Calendar variations in the phenomena of nature and the apparition of two Higgs bosons, http://www.bestthinking.com/articles/science/physics/quantum_physics/calendar-variations-in-the-phenomena-of-nature-and-the-apparition-of-two-higgs-bosons

24. S. Berkovich, A comprehensive explanation of quantum mechanics. www.cs.gwu.edu/research/technical-report/170 http://www.bestthinking.com/articles/science/physics/quantum_physics/a-comprehensive-explanation-of-quantum-mechanics

25. P.L. Kapitsa, The future of science, in *Experiment, Theory, Practice* (Publishing House "Science", Moscow, 1987), A.S.Borovik-Romanov and P.E.Rubinin. (Eds) pp. 395–418 (in Russian)
26. L. Grossman, Fireworks erupt over quantum nonsense. New Sci. 22 September 2012, 13 (2012)
27. S. Clark, Half-life heresy: seasons change in the atom's heart. New Sci. 14 November 2012(2891), 42–45 (2012)
28. B.G. Schwartz, R.A. Kloner, Abstract 11723: Seasonal variation in cardiac death rates is uniform across different climates, http://circ.ahajournals.org/cgi/content/meeting_abstract/126/21_MeetingAbstracts/A1172

Alert Data Aggregation and Transmission Prioritization over Mobile Networks

Hasan Cam, Pierre A. Mouallem and Robinson E. Pino

1 Introduction

Network-based and host-based intrusion detection systems (IDS) in wireless mobile networks rely on the ability of mobile nodes to monitor transmission activities of each other and analyze packet contents to detect intrusions. However it is shown experimentally that these types of IDSs generate high amounts of false alarms [1, 2], leading to poor intrusion detection performance and affecting adversely the already bandwidth-limited communication medium of wireless mobile networks. Therefore, alert aggregation is needed to reduce the amount of alerts without losing important information, while improving accuracy of decision making process. This work presents a multi-stage real-time alert aggregation technique for mobile networks that greatly reduces the amount of alert and data transmitted and attempts to maximize the bandwidth utilization. The goal is to transmit the alerts generated by all nodes to a central repository, like Forward Operating Base (FOB), where they are processed, analyzed and a feedback is provided accordingly.

This technique uses alert attributes, such as source IP addresses, to create a number of alert attribute sets where each set is constituted of one or more attributes. As raw alerts are generated, they are aggregated into meta-alerts according to an attribute set. The meta-alerts of each set are kept in a separate queue. A probabilistic distribution function is then calculated for each queue to determine the importance of meta-alerts within, which also dictates the queue that gets transmitted first. If a node is within range of the FOB, it transmits its meta-alerts

H. Cam (✉) · P. A. Mouallem · R. E. Pino
Network Science Division, US Army Research Laboratory, Adelphi, MD 20783, USA
e-mail: hasan.cam.civ@mail.mil

P. A. Mouallem
e-mail: pierre.a.mouallem.ctr@mail.mil

R. E. Pino
e-mail: robinson.e.pino.ctr@mail.mil

R. E. Pino (ed.), *Network Science and Cybersecurity*,
Advances in Information Security 55, DOI: 10.1007/978-1-4614-7597-2_13,
© Springer Science+Business Media New York 2014

directly to it. However if a node is not within range, then it would transmit its meta-alerts to an adjacent node that is closest to the FOB. That node would then aggregate the meta-alerts it receives from adjacent nodes with its own meta-alerts before it forwards the results. This work greatly reduces the amount of bandwidth required to transmit the alerts. It also prioritizes the alerts for transmission so that important alerts get transmitted first.

The rest of this chapter is organized as follows. Section 2 discusses the related work on data aggregation. Section 3 present the proposed data aggregation and prioritization techniques. Section 4 describes the experimental setup and results. Section 5 concludes with final remarks and future work.

2 Related Work

In order to mitigate the adverse impact of high rates of intrusion detection systems alerts, several alert aggregation techniques [3–6] have been proposed in the literature to simplify and aid in the analysis of such alerts. However most of those techniques focus on wired networks, where bandwidth and distance is not usually a concern. Relatively fewer research efforts have been made to mobile ad-hoc networks based IDS. Some of that research includes consensus voting among mobile nodes [7], adding additional components to detect routing misbehavior [8], or a combination of the two [9]. The technique presented in this chapter takes advantage of consensus voting among nodes but further extends it by aggregating alerts across neighboring nodes. It also prioritizes the aggregated alerts based on their criticality and takes into account alert patterns and feedback from analysts.

Alert aggregation in general encompasses techniques such as normalization [4], correlation [5] and aggregation [3, 6]. Normalization centers on the "translation" of the alerts that originate from different intrusion detection systems and sensors into a standardized format that is understood by analysis components. There has been some work done towards achieving a common representation of alerts [4], however there are still several gaps that faces normalization. The most significant gap is the lack of a common taxonomy of attacks, resulting in each IDS classifying alerts differently. Some effort have been made to address this gap [10, 11], but it is still far from wide acceptance.

Correlation centers on finding the relationships between alerts in order to reconstruct attack scenarios from isolated alerts. Correlation aims at providing an abstraction of raw alerts, giving analyst a higher level view of attack instances. However, correlation is not as effective at reducing the number of the raw alerts. There are several correlation techniques and they have been extensively covered in the literature [12, 13].

Aggregation centers on grouping alerts with common parameters. Alerts can be grouped based on shared attributes [14], root causes [15] or a combination of the two. Several techniques have been used for alert aggregation, such as data mining, machine learning, and neural network. The work presented in this chapter will focus mainly on the aggregation technique as it offers the highest level of alert reduction among the three different techniques.

Additionally, work has been done in the area of alert prioritization, which aims at classifying alerts based on their severity/importance [16]. The goal is to help analysts or systems administrators to take appropriate actions based on the class of alerts. However the work available in the literature is somewhat limited and still lack maturity. This chapter also tackles this topic and presents a prioritization scheme to maximize bandwidth efficiency.

3 Alert Aggregation Technique

This section presents our alert aggregation technique. We begin by describing the network topology in which this technique operates, and then present the proposed technique. We then discuss the implementation of each step in a distinct subsection using algorithms and providing examples.

Protocol Alert Aggregation
Input: locally generated alerts
Output: Meta-Alerts and Feedback.

1. Aggregate locally generated alerts
2. If non-local meta-alerts are received from other nodes, aggregate them with local meta-alerts
3. Prioritize and prepare meta-alerts for transmission
4. When transmission window is available, transmit meta-alerts to FOB if within range, or to an adjacent node that is closest to FOB
5. FOB receives meta-alerts and processes them
6. Feedback is formulated by the FOB and propagated back to the nodes.

Figure 1 shows the timing diagram of the protocol above for a node N_k. Note that for nodes that are within one hop of the FOB, timing t_2–t_3 would be zero.

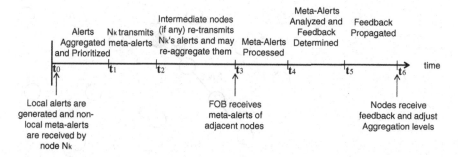

Fig. 1 Timing for protocol alert aggregation

3.1 Network Topology

Figure 2 shows an example of the network topology in which this technique shall operate. It is comprised of mobile nodes, each running an IDS, and a forward operating base (FOB), also known as base station, that receives the alerts from the nodes and analyzes them.

Each node aggregates the alerts it generates/receives and sends the resulting meta-alerts to an adjacent node with a lower hop count than the current node. The nodes that are within one hop of the FOB sends their meta-alerts directly to the FOB. The FOB would then process and analyzes the received meta-alerts and provide feedback to the nodes.

Next, we describe the Alert aggregation algorithm for local and non-local alerts.

3.2 Local Alert Aggregation

The most significant alerts attributes are usually considered to be source and destination (i.e., target) IP addresses, and signatures or anomaly/behavioral patterns, depending on whether IDS is based on signatures or anomaly/behavioral patterns. The other attributes may include port numbers, event detection time-stamp, event duration, TCP status flags, service type, and size of transmitted data. Alerts can be classified as categorical (e.g., IP addresses, port numbers) and continuous (e.g., duration).

In the proposed algorithm, a number of alert attribute sets are formed such that each set is constituted by one or more alert attributes, and an attribute can be a

Fig. 2 Network topology
(add a bit more information
about figure)

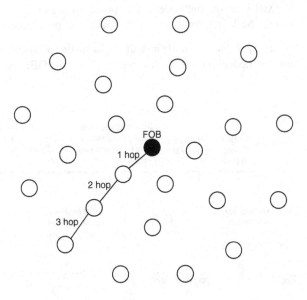

member of more than one set. Some examples for sets of attributes could be: {signatureID}, {signatureID, srcIP}, {signatureID, srcIP, portID}, {signatureID, duration}, {signatureID, srcIP}, and {signatureID, srcIP, dstIP, portID}. The meta-alerts of each set are kept in a separate queue. As for the meta-alerts parameters that are not part of the attribute set, they are stored as a list within the same meta-alert.

The proposed algorithm denotes a raw alert by a 7-tuple: raw-alert ID, source IP address (srcIP), destination IP address (dstIP), signature type or ID (sigID), Port ID (portID), alert event occurrence time (T), node ID and log data (logdata).

Similarly, the aggregated alerts, called meta-alerts, are denoted as follows: meta-alert ID number, number of raw alerts represented by a particular meta alert (N), time of meta-alert updating (Te), aggregate attribute 1 through n, and a list of tuples holding the raw alert parameters not used in the aggregation.

Figure 3 shows an example of Alert Aggregation. It shows the aggregation of three raw alerts generated at a single node into one meta-alert based on two attributes, srcIP and SigID.

Each attribute set is represented by a meta-alert queue. Attributes can be used in multiple queues, for example the attribute set of queue one can be {SigID}, and the attribute of queue two can be {SigID, SrcIP}.

The local alerts aggregation algorithm is shown next. It takes as input raw alerts and output meta-alerts to be send to the next node.

Algorithm: Local Alert Aggregation
Input: Raw Alerts
Output: Meta-Alerts (grouped in queues)

1. *Initialize Attribute Sets*
2. *Initialize meta-alerts queues*
3. *For each alert read:*
       If queues empty:
           then
               Randomly assign Alert to queue and generate meta-alert
         else
               Check Queue with highest attribute count,
               if Alert can be aggregated:
               then
                   Aggregate Alert
               else
                   Check remaining queues,
                   If Alert can be aggregated:
                   then
                       Aggregate Alert
                   else
                     Randomly assign Alert to queue and generate new meta-alert

---

Upon the receipt of a new raw alert, we check if the queues are empty, if they are, the raw alert is assigned to a random queue and a Meta-Alert is created containing that alert.

Raw Alert 1	AlertID1	Src IP1	Dst IP1	Sig ID1	Port ID1	T1	Node1	Log Data1

Raw Alert 2	AlertID2	Src IP1	Dst IP2	Sig ID1	Port ID2	T2	Node1	Log Data2

Raw Alert 3	AlertID3	Src IP1	Dst IP2	Sig ID1	Port ID3	T3	Node1	Log Data3

Alert Aggregation on {SrcIP,SigID}

Meta-Alert 1	MA ID	MA Te	Number of Raw Alerts	Src IP	Sig ID	{ID1, dstIP1, portID1, T1, Node1, logdata1} {ID2, dstIP2, portID2, T2, Node1, logdata2} {ID3, dstIP2, portID3, T3, Node1, logdata3}

Fig. 3 Alert aggregation example

For each subsequent raw alert received, we check if the alert can be aggregated with existing meta-alerts, starting with the queues that have the highest number of attributes. If it can, then the alert is aggregated, if not, then we check the remaining queues for possible aggregation. If no aggregation is possible, and then the alert is assigned to a random queue and a new meta-alert is created containing the received alert.

Figure 4 further shows the steps taken when a new raw alert is received.

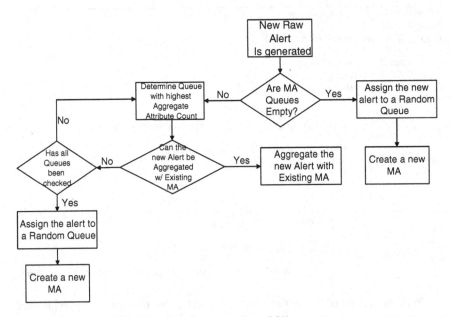

Fig. 4 Aggregation of local raw alerts into meta-alerts (MA)

3.3 Non-local Alert Aggregation

In addition to handling local alerts, each node is capable of receiving meta-alerts from adjacent nodes. The non-local meta-alerts are aggregated with local meta-alerts when possible, and the resulting meta-alerts are forwarded to the next node. The non-local meta-alerts aggregation algorithm is shown next. It takes as input the meta-alerts received from other nodes, and adds them to existing queues or generates new queues.

Algorithm: non-local meta-alerts Aggregation.
Input: Non-local meta-alerts queues
Output: none

1. For each non-local meta-alert queue received:
    Compare its attribute set to the attribute sets of local meta-alert queues,
    If attribute set match:
        then
            Aggregate its meta-alerts with existing meta-alerts in local queue
            Discard non-local meta-alerts queue
        else
            Store non-local meta-alerts queue as local

Figure 5 shows the handling of non-local meta-alerts.

3.4 Alert Prioritization

Since bandwidth is limited in the type of deployments, alerts need to be prioritized so that the most critical alerts get transmitted first once a transmission window becomes available. Alert prioritization is achieved by calculating the probability distribution for each queue. The meta-alerts of the queue with the highest value are

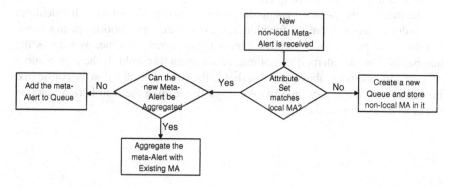

Fig. 5 Handling of non-local meta-alerts

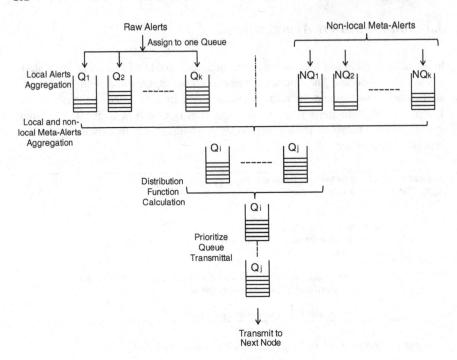

Fig. 6 Aggregation of raw alerts and aggregation of local and non-local meta-alerts for a single node

chosen to be forwarded to the next node in the network (or to the FOB in the case of the nodes connected to the FOB).

Additionally, each node that receives meta-alerts from other nodes will determine which meta-alerts set to forward first based on the value of their polynomial distribution functions.

Figure 6 shows an example for a single node of the aggregation of generated alerts, and the aggregation of local and non-local meta-alerts. It also shows the transmission prioritization of queues.

The meta-alerts queues transmission prioritization is shown next. It calculates the multinomial distribution function for each queue and prioritizes the transmission. The prioritization expires after a preset period of time, as long as the number of new raw alerts does not exceed a certain threshold. If the new number of new raw alerts is above the threshold, then the prioritization is considered invalid even if the expiry time has not elapsed, meaning the distribution function needs to be recalculated.

Algorithm: meta-alerts queues prioritization.
Input: meta-alerts queues
Output: none

1. *Check prior multinomial distribution function value for each queue*
   If value exists and has not expired:
   then

     Read the number of incoming alerts since the last calculation
     If the number above the pre-set threshold:
     then

       Calculate multinomial distribution function
     else

       Prior value still valid, no need to re-calculate
   else

     Calculate multinomial distribution function

2. *Prioritize transmission order based on calculated value*

3.4.1 Distribution Functions Calculation

To aggregate alerts based on the alerts attributes or parameters, our technique assumes: (1) an attack instance is a random process generating alerts, and (2) attributes of alerts have different probability distributions.

We compute (1) categorical attributes have multinomial distributions; (2) continuous attributes have standard normal distributions. Note that multinomial distribution of an alert with a categorical attribute set is equal to the product of probabilities of all event(s) that lead to the alert generation. If an alert has two categorical attributes, say signature ID and source IP address, then its multinomial distribution equals the product of their probabilities. In a multinomial distribution, each trial results in one of some fixed finite number k of possible outcomes, with probabilities $p_1, ..., p_k$.

The multinomial distribution can be calculated using a probability mass function, with the following equation:

$$P(Q_j) = \frac{N!}{(n_1! n_2! \ldots n_k!)} \prod_{i=1}^{k} p_i^{n_i} \tag{1}$$

where
Q_j is meta-queue j
N is the number of raw alerts
k is the number meta-alerts in Q
n_i is the number of alerts aggregated in meta-alert i
p_i is the probability of observing meta-alert i

Additionally, p_i can be calculated as follows:

$$p_i = W_{A_i} \times W_{IAi} \times \frac{n_i}{N} \tag{2}$$

where

Pi　　is the probability of meta-alert i
W_{Ai}　is the weight of attributes of the queue containing meta-alert i
W_{IAi}　is the weight of the values of attributes of meta-alert i
n_i　　is the number of alerts in meta-alert i
N　　is the number of raw alerts

The weights of the attributes and the weights of the values of the attributes are calculated using a combination of historical data, analysts feedback and incident reports.

As far as normal distribution, it can be calculated using the following probability density function:

$$\phi(x) = \frac{1}{\sqrt{2\pi}} e^{-\frac{1}{2}x} \tag{3}$$

3.5 Alert Transmission

As bandwidth is limited in this type of deployments, Meta-Alerts are transmitted when bandwidth becomes available. The alert transmission algorithm is shown next. It takes as input the ordered meta-alerts queues and transmits them based on the availability of transmission windows. Before transmittal, it checks if the meta-alert timer has not expired, and that the raw alerts contained within that meta-alert have not been previously transmitted within other meta-alerts. If they haven't, they are transmitted, if they have, then they are stripped out and discarded.

```
While transmission window is open:
    check meta-alert time
    If meta-alert time has expired:
        Then
                Discard alert
        else
                Check transmitted alerts list,
                If alert has been transmitted:
                Then
                        Discard alert
                else
                        Transmit alert
```

3.6 Feedback Propagation

Based on the analysis of received meta-alerts, the analyst at the FOB might choose to give particular attention to certain alerts, or might choose to increase the amounts of raw packet data received that pertain to certain alerts. To achieve that, a feedback transmission mechanism that uses multicast is proposed. The FOB would multicast its feedback preference to all connected nodes, and those nodes would in turn forward that message to their connected nodes. Simple routing algorithms are used to avoid routing loops.

4 Experimental Validation

4.1 Experiment Setup

To evaluate the algorithms presented, 2 weeks' worth of data were collected across ten nodes, from October 1st 2012 till October 14 2012, from an enterprise-scale production network intrusion detection system (NIDS), using a combination of Snort Emerging Threats (ET) and Vulnerability Research Team (VRT) rule sets. Figure 7 shows the nodes inter-connectivity and the connection to the Forward Operating Base (FOB). Alerts flow from the outer most nodes towards the FOB. For simplicity we assume that no routing loops exist and that each node transmits to a single node.

To perform the Data Aggregation at each node, we use the same attribute sets across all nodes, with each attribute set corresponding to a meta-alert Queue. Note that the attribute sets are not set in stone and can be defined/modified based on the analysis of previously collected data and feedback from analysts. The attribute sets used are shown in Table 1.

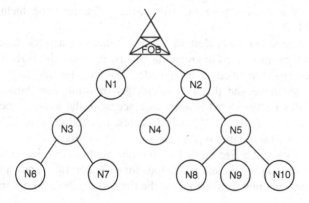

Fig. 7 Field nodes connectivity

Table 1 Attribute sets

Attribute set	Description
{AlertID}	IDS alert ID
{AlertID, srcIP}	IDS alert ID, source IP address
{AlertID, srcIP, portID}	IDS alert ID, source IP address, destination port ID
{AlertID, srcIP, portID, dstIP}	IDS alert ID, source IP address, destination port ID, destination IP address
{AlertID, duration}	IDS alert ID, duration of alert

Table 2 Raw alerts count per node

Node ID	Raw alerts
1	339,311
2	148,320
3	216,802
4	228,448
5	307,169
6	33,660
7	43,326
8	22,779
9	111,707
10	119
Total	1,451,633

The total number of alerts collected was approximately 1.5 million across the ten nodes. Table 2 shows the number of alerts collected at each node.

4.2 Experiment Setup

We first run the aggregation algorithm locally at each node. The raw alerts are aggregated and divided among the five queues. The resulting meta-alerts are shown in Table 3.

The total number of meta-alert in Table 3 represents a worst case scenario, when no non-local meta-alert aggregation occurs, meaning the FOB will receive all the meta-alerts generated at each node. This can be due to the fact that bandwidth is available and that meta-alerts get transmitted without any delay. Even though this is considered a worst case scenario, the total number of meta-alerts received by the FOB is 1,109, which represents a reduction of 99.92 % when compared to the total number of raw alerts ($\sim$ 1.5 million).

Next we compare the size on disk of the raw alerts to the size of meta-alerts. Additionally we compute the size of the logs for each alert (also known as trim file, which represents the raw data stream for the time where the alert occurred). Since

Table 3 Local alert aggregation per nodes

Node ID	Raw alerts	Meta-alerts	Queue 0	Queue 1	Queue 2	Queue 3	Queue 4
1	339,311	129	29	18	19	30	33
2	148,320	93	15	21	14	23	20
3	216,802	43	14	7	5	5	12
4	228,448	165	32	34	33	39	27
5	307,169	169	30	38	33	39	29
6	33,660	350	71	62	71	77	69
7	43,326	70	13	15	24	10	8
8	22,779	35	8	6	6	7	8
9	111,707	48	8	8	11	13	8
10	119	7	2	1	2	1	2
Total	1,451,633	1,109	222	210	218	244	216
Reduction		99.92 %					

the log files cannot be compressed or reduced, we consider three alternatives to transmitting those logs:

1. Transmit all logs for all alerts.
2. Transmit the logs for the first alert only in a given meta-alert.
3. Do not transmit any log files, instead store them locally.

Table 4 summarizes the results observed.

A reduction of 57.14 % is realized for the alerts disk size (alerts only). As for total size reduction, for alternative 1, the reduction is 7.96 %, which is expected, since the raw logs consume a significant amount of disk space when compared to alerts. For Alternative 2, the reduction is 92.57 %, which is a great improvement over alternative 1, and most likely to be used among the three alternatives.

As mentioned earlier, the results presented in Table 3 would be the worst case results since no non-local meta-alert aggregation was done. The best case scenario would be when each node receives all possible non-local meta-alerts before it transmits its own meta-alerts, which means each node would have an opportunity to run the non-local meta-alert aggregation algorithm, resulting in further compression. Table 5 shows the results of simulating this scenario, where nodes N3, N4 and N5 do not transmit until they receive the meta-alerts from nodes N7, N8,

Table 4 Size (in Kb) of raw and meta-alerts

	Amount	Size of alerts	Log (trim) files size	Total size	Reduction %
Raw alerts	1,451,633	8,876	54,808	63,684	–
Meta-alerts with full log files	1,109	3,805	54,808	58,612	7.96
Meta-alerts with partial log files	1,109	3,805	929	4,734	92.57
Meta-alerts with no log files	1,109	3,805	–	3,805	94.03

Table 5 Local and non-local meta-alert aggregation (best case scenario)

Node ID	Local MA	Received non-local MA	Aggregated local and non-local MA	Total MA received at the FOB
1	129	412	445	710
2	93	352	390	
3	43	420	412	
4	165	–	165	
5	169	90	187	
6	350	–	350	
7	70	–	70	
8	35	–	35	
9	48	–	48	
10	7	–	7	

N9 and N10. Similarly nodes N1 and N2 do not transmit until they received the meta-alerts from nodes N4 and N5.

As we can observe in the table above, the total number of meta-alerts received at the FOB in the best case scenario is 710, which represents a reduction of 35.98 % when compared to the total number of meta alerts noted in Table 3, and a reduction of 99.95 % when compared to raw alerts (1.5 million).

Furthermore, the size of the meta-alerts received by the FOB is 3,120 Kb, which is a reduction of 18 % when compared to the size of meta-alerts recorded in Table 4. Table 6 summarizes the results observed.

A reduction of 64.85 % is realized for the alerts disk size (alerts only). As for total size reduction, for alternative 1, the reduction is 9.04 %, which is expected, since the raw logs consume a significant amount of disk space when compared to alerts. For Alternative 2, the reduction is 93.89 %, which is a great improvement over alternative 1, and the most likely to be used among the three alternatives.

The following figure shows the size of meta-alerts (in blue) and the corresponding data for those meta-alerts (in red) that are transmitted by each node. Note that nodes 1, 2, 3 and 5 transmit a higher amount of data since they receive meta-alerts from other nodes and aggregate those alerts with the local meta-alerts (Fig. 8).

Table 6 Size (in Kb) of local and non-local meta-alerts

	Amount	Size of alerts	Log (trim) files size	Total size	Reduction %
Raw alerts	1,451,633	8,876	54,808	63,684	–
Meta-alerts with full log files	710	3,120	54,808	57,928	9.04
Meta-alerts with partial log files	710	3,120	769	3,889	93.89
Meta-alerts with no log files	710	3,120	–	3,120	95.10

Fig. 8 Size (in Kbps) of meta-alerts and their data per node

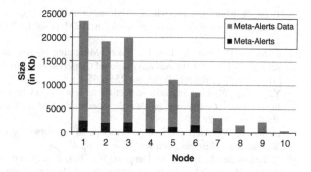

5 Conclusion and Future Work

This book chapter has presented a new alert data aggregation technique based on alerts attribute sets for alerts generated by intrusion detection systems in mobile networks. It has also presented techniques for alert transmission and prioritization, and feedback control. We've also simulated our aggregation techniques using real alerts captured from an enterprise network and the results show that our techniques can achieve a very high percentage of aggregation and adapt to the available bandwidth to maximize its efficiency.

In this chapter, we do not consider computing power or the energy available at each node. Taking them into consideration might affect how often we can prioritize meta-alerts since the prioritization process is a computing intensive operation. Future work will include further analysis of those additional parameters in order to optimize alert aggregation and energy consumption. Future work also includes further examination of the utilization, and how it might affect our aggregation models, and how risk is assessed based on the chosen models.

References

1. F. Cuppens, Managing alerts in a multi-intrusion detection environment, in *Proceedings of the 17th Annual Computer Security Applications Conference, 2001*
2. H. Farhadi, M. Amirhaeri, M. Khansari, Alert correlation and prediction using data mining and HMM. ISC Int. J. Inform. Secur. **3** (2), (2011), 1–25
3. R. Sadoddin, A. Ghorbani, Alert correlation survey: framework and techniques, in *PST'06: Proceedings of the 2006 International Conference on Privacy, Security and Trust*, New York, 2006, pp. 1–10
4. D. Curry, H. Debar, Intrusion detection working group: Intrusion detection message exchange format data model and extensible markup language (xml) document type definition. RFC 4765, Internet-Draft (2003): 21–26
5. F. Cuppens, A. Miege, Alert correlation in a cooperative intrusion detection framework, in *SP '02: Proceedings of the 2002 IEEE Symposium on Security and Privacy*, Washington, 2002, p. 202

6. T. He, B. Blum, J. Stankovic, T. Abdelzaher, AIDA: Adaptive application-independent data aggregation in wireless sensor networks, in ACM Trans. Embedded Comput. Syst 3.2 (2004): 426–457

7. S. Marti, T. Giuli, K. Lai, M. Baker, Mitigating routing misbehavior in mobile ad hoc networks, in *The 6th Annual International Conference on Mobile Computing and Networking (MobiCom'00)*, Boston, 2000, pp. 255–265

8. Y. Zhang, W. Lee, intrusion detection in wireless ad hoc networks, in *The 6th Annual International Conference on Mobile Computing and Networking (MobiCom'00)*, Boston, 2000, pp. 275–283

9. B. Sun, K. Wu, Alert aggregation in mobile ad hoc networks, in *Proceedings of the 2nd ACM Workshop on Wireless Security, Wise 03*, pp 69–78

10. C. Landwehr, A. Bull, J. McDermott, W. Choi, A taxonomy of computer program security flaws, with examples, Technical report, Naval Research Laboratory, Nov. 1993

11. U. Lindqvist, E. Jonsson, How to systematically classify computer security intrusions, in *IEEE Symposium on Security and Privacy* (1997), pp. 154–163

12. F. Cuppens, A. Miege, Alert correlation in a cooperative intrusion detection framework, in *SP '02: Proceedings of the 2002 IEEE Symposium on Security and Privacy*, Washington, 2002, p. 202

13. L. Wang, A. Ghorbani, Y. Li, Automatic multi-step attack pattern discovering. Int. J. Netw. Secur. **10**(2), 142–152 (2010)

14. E Valdes, K. Skinner, Probabilistic alert correlation. Lect. Notes Comput. Sci. **2212**, 54–68 (2001)

15. K. Julisch, Clustering intrusion detection alarms to support root cause analysis. ACM Trans. Inf. Syst. Secur. **6**(4), 443–471 (2003)

16. B. Morin, L. Me, H. Debar, M. Ducasse, M2d2: a formal data model for ids alert correlation, in *RAID* (2002), 115–127

Semantic Features from Web-Traffic Streams

Steve Hutchinson

1 Objective

Techniques and processes for semantic analysis of prosaic documents are well established. Indeed, numerous probabilistic, generative models have been developed and are shown to adequately model somewhat complex relationships between authors, links/references/citations, hierarchy of topics, bi-gram ordering of tokens and other interesting relationships. Semantic analysis has also effectively been applied to music genre similarity matching, and to other types of media such as streams of Twitter messages. Some of the issues arising when considering a stream as a corpus have been addressed in the works of Wang et al. [1, 11].

In cyber-security, there is a growing requirement to conduct meaningful semantic analysis of streams, but, in this case, applied to online browsing sessions. This application area is significantly different from traditional document classification with semantic analysis in that:

- The corpus does not have a unifying domain or feature that relates all corpus members to be considered as a batch of documents.
- The corpus is comprised of weakly or irregularly delimited 'pseudo-documents' which may also be nested or interleaved during capture and chronologically ordered.
- The semantics in which we have interest pertain to the pairing of an information retrieval request and its responses. We combine the request and responses together to form a document. We are more interested in the semantics of the session as informed by semantic analysis of the constituent documents observed.
- The documents in each such corpus were neither authored by, nor classified or grouped by the requester.

S. Hutchinson (✉)
ICF International, Fairfax, VA, USA
e-mail: Steve.Hutchinson@icfi.com

R. E. Pino (ed.), *Network Science and Cybersecurity*,
Advances in Information Security 55, DOI: 10.1007/978-1-4614-7597-2_14,
© Springer Science+Business Media New York 2014

This paper focuses on a segmentation and tokenization process to prepare for semantic and topic analysis. The resulting output from the process can be used to form semantic feature vectors for subsequent machine-learning and profiling processes. These latter processes are not described here.

2 Processing Web-Traffic as Corpora

A significant issue when dealing with online traffic as documents is related to the difference between a fixed corpus and a stream or session. We need a way to track topic-words between corpora and observed sessions. Since in traditional methods, semantic topics are usually determined by probabilistic analysis of numeric vector-space document vectors that represent term (word token) frequencies, we would like each TermID to represent the same word throughout the entire lifetime of obser-vations as well as across users (represented and anonymized by IP address). This suggests that a pre-defined lexicon should form the basis for word tokenization.

Another significant issue in semantic analysis concerns the recognition and resolution of synonyms. Although not usually a concern in traditional semantic analysis, we anticipate analysis scenarios where domain-specific synonym set creation and resolution will significantly improve semantic analysis of these streams. This notion is of course dependent upon a lexicon-based approach.

In order to make use of the extant computational methods for semantic analysis, it is desirable to map the characteristics of web traffic into formats compatible with these methods. Thus, a TermID:Frequency (T:F) document matrix representation of these web pseudo documents will allow analysis of blocks or sets of documents using existing tools. A (T:F) representation is similar to the classic TF-IDF (term-frequency inverse document frequency) and is widely supported by document classification and semantic analysis tools.

3 Related Work

In linguistic analysis, segmentation is the process of partitioning a content corpus into smaller segments, usually into 'documents' and 'sentences' within documents. Segmentation is greatly facilitated by the presence of punctuation in written and electronic corpora; such punctuation (such as the period, semicolon, or other characters) often faithfully represent the author's intent to delimit such clauses and sentences. Similarly, segmentation into documents is often facilitated by the original structural organization of the corpus; each document is often contained within a separate file in a file system, or is delimited by an end-of-file marker in a multi-document archive.

In other forms of discourse, especially in conversations, transcriptions, and streaming/media, sentence- and document-delimiters are seldom available. Others have proposed solutions (albeit often application-specific) to segmentation. Hearst [2] describes the method of 'TextTiling' in which the goal is to annotate each document with appropriate sub-structure. Principle clues of sub-structure are obtained via the presence of 'headings' and 'subheadings' to divide the text. In the absence of these heading markers, TextTiling implements an algorithm for lexical cohesion to partition the text document into multi-paragraph segments reflecting a subtopic structure.

Jain [3] proposes a method to divide documents into segments based upon topic and subtopics. Such labeled subsections or paragraphs would facilitate analyses and retrieval by a semantic query. This technique employs the TextTiling of [2] using a sliding window. Similarity scores between adjacent groups of sentences allow groupings of sentences by similarity, as an indicator of topic.

Kern [4, 9] observes that there is a difference between linear text segmentation (splitting a long text into chunks of consecutive text blocks or sentences) and hierarchical text segmentation to split documents into finer grained, topic-segments. They also observe that the motivation for authors to insert a paragraph break is not identical to the topic shift detected by lexical cohesion and other algorithms. Kern proposes a method, TSF ('TextSegFault'), which uses a sliding window approach to identify topic boundaries by parametric similarity and dissimilarity measures between blocks within sentences.

4 Our Purpose for Segmentation

The above cited methods for segmentation operate by performing delimiter-based, or content sentence coherence algorithms on the original, ordered content. Many of the above methods purpose to annotate or label each paragraph or group of sentences with a prevailing topic semantic for subsequent query or analysis. For our proposed (future) analysis of web-traffic sessions, we use a different model for documents. Here, we consider a document to commence with an observed HTTP request and consist of all the assembled response content received for this (each) request. This approach has two principle characteristics: (1) it allows the content of the request string to be associated with the resulting response text—for use in document classification studies; (2) it results in each document being expressed by a bag of tokens with no other discernible structure.

5 Tokenization

In linguistic analysis, tokenization is the method used to assign a unique, integer identifier, the TermID, to each distinct character string of interest. Documents can thus be represented as un-ordered, or ordered sequences of tokens, by their TermIDs. It is customary to remove undesired strings during this process; this is often accomplished through use of a stop-word lexicon. Stemming is another typical process used to regularize related strings into one, common representation and TermID. A common stemming algorithm is the 'Porter Stemmer' developed by Martin Porter in 1980 [5]. The above methods are natural-language specific, that is semantic analysis and tokenization are driven by language-specific lexicons, and stemming algorithms when appropriate.

Futrelle [6] develops a method called Extreme Tokenization, to apply to HTML documents in the biology literature domain. In this method, papers (documents) are converted to a sequence of tokens (integer Token IDs or TIDs) including tokens for the HTML markup, and labeled with a generated Sequence ID (SID) which functions as the accession number for each Archival Token Sequence (ATS). White space is reduced to a single blank-space token. Entities and other contiguous sequences of characters [a-zA-Z0-9] are also retained. A lexicon is constructed after tokenization which maps between TIDs and strings. Other specialized lexicons are generated to identify capitalizations, delimiter cases (full stop, comma, etc.), so as to avoid re-parsing during search, display, and sentence boundary detection. Documents thus processed are stored as XML documents for compatibility with databases, storage containers, and document processing query systems.

6 Motivation for Future Exploration

Our motivation for this work in segmentation and tokenization to form a novel corpus type for web-traffic is the desire to form a time-series of semantic spectra and support two analyses: (1) longitudinal spectral evolution—modeling similarities and differences of spectra over some longer time period to identify repeating topic 'spectral lines' that characterize a workstation (user's) browsing; and (2) a differential comparison technique to measure the difference between two sets of spectra, or possibly to match an observed spectrum to an existing archived set. The analogue between emission-spectra in quantum mechanics and semantic analysis has been studied in Wittek [7, 10] although for a different purpose.

7 Implementation

To implement segmentation and tokenization appropriate for semantic analysis of web-traffic, we amend the processes above with these additional functions: (1) we use GET-response sets as the means to define and segment traffic into documents, (2) construe all web-browser traffic sessions as a set and use this sequence of documents forming a windowed-corpus for subsequent analysis, (3) include the string-literal contents of the GET request with the response content, forming a document type more compatible with existing document clustering and classification analysis, (4) define a new type of document corpus with its members not associated by a common domain or attribute set, but rather associated by the utilization intent of a specific user, (5) and use a lexicon to generate canonical representations of topic semantics for comparison and aggregation across corpora (windowed corpora).

The detailed steps of this process are shown below:

Process step	Description
Traffic collection	A sensor with visibility into the subnet containing the IP addresses of interest collects streams of (web) traffic—essentially all session flows that contain the IP of interest either as a source or destination. The collection tool should capture entire packets, rather than truncating (such tools often truncate at 53 bytes by default)
Segmentation	Formation of 'documents'. Each document commences with a GET/POST request, accumulates all responses for the immediately enclosing GET, and terminated by the next GET/POST
Tokenization: structural cleaning to tokens	Removal of punctuation and delimiters, replacing each with one space to preserve token integrity. After cleaning, tokens are the remaining strings of printable characters delimited by spaces
Tokenization: stop-word removal	Removes tokens that appear in a crafted stop-word list, appropriate for this type of web-traffic. Try to remove key-tokens of key-value pairs (KVPs)
Tokenization: lexical mapping and (future) synonym resolution	Although lexicon TermID mapping, could be applied after semantic analysis by mapping the resulting topic-words into a standard lexicon, we choose to perform this mapping beforehand to reflect a (domain specific) resolution of synonyms. This step also forms an IP-address specific auxiliary lexicon, allowing it to be observed and managed separately
Document matrix: canonical representation with a chosen lexicon	Use a static lexicon when converting tokens to TermIDs so that the same word, and hence the same topic semantics are retained across analysis batches, and between different user-sessions
Formation of term: frequency document vectors	Using TermIDs from the standard lexicon. Create an auxiliary lexicon to contain IP-specific entities that have semantic and differential value
Semantic analysis	Choose a standard number of topics (N). Non-parametric analysis could be performed and considered in the future. In this study, we chose topic size of 10 with each expressing a vector of the top 10 words for each topic
	Use a standard method on the resulting corpus of web documents. Perform threshold or ranking selection for the size of the set of tokens representing each topic
Semantic feature vectors	Identified by IP address and time-range. Archiving and machine learning techniques for trend and anomaly analysis

In the above multi-step process to perform lexicon-driven tokenization of web-traffic streams, the first step is to segment the observed traffic into sessions. Typically in cyber-security, we form collected sets of observed traffic to and from one selected IP address (at a time). Streams are further segmented into non-overlapping batches of flows during an observation period. (It is customary to accumulate traffic into hour-long sets for subsequent processing and analysis). Lastly, sessions are segmented into pseudo documents by combining each GET/POST request with related response traffic; this pair constitutes a document.

Tokenization implements lexical parsing to separate words from other content strings and characters. Through tokenization, we endeavor to make web-traffic resemble a traditional document by:

- Translation of domain-specific delimiters into appropriate token strings. As an example, in web-traffic, the forward-slash acts as a delimiter in the URL path component. Also, in a URL, the query character ('?') delimits the path/file component from clauses of the query. The equals and ampersand characters ('=', '&') are used to delimit key-value pairs in a URL.
- Translation of punctuation to a standard delimiter (space). After special character delimiters are identified, the resulting tokens are delimited by a space character. (In the future, we will retain key-value pairs for special processing during tokenization. The value components are often related to the 'semantics' of the enclosing web-page, context, and container, and can be used in semantic analysis, whereas the key string components are often scripting variable names and query keywords which are often of lesser interest).
- Selective preservation of numeric and special characters. In many instances, mixed case, mixed alpha-numeric, and the URL components described above have semantic significance and thus are preserved in our tokenization process.

After tokenization, the resulting output is a set of delimited strings, where a string is a space-delimited, contiguous sequence of printable characters.

Finally, there are certain strings or tokens we treat as a 'word' if that identical string is found in our selected lexicon. In traditional document analysis, stemming is performed to convert alternate forms into a canonical form. For example, the words 'walking', 'walks', and 'walked' would be converted to the single stem: 'walk'. We maintain that stemming is not appropriate for our intended treatment of web-traffic, because we are interested in the dispersion, frequency and implied semantics of the distinct forms of words. This is consistent with our similar treatment of named entities, as well as other alpha-numeric tokens when it is likely they are being used in a semantically interesting manner (that is, they are not being used to delimit or compose containers themselves).

8 Concepts for Processing of Web-Traffic Streams

Most of the concepts and processes we employ for web-traffic streams are well understood and precise. Since our process extends some of these notions to surmount particular problems inherent to web-traffic, we restate some of these concepts and explain those aspects we deem particularly important. Other concepts we extend for particular situations. We also readily acknowledge that while certain of these concepts involving lexicons and stop-word lists have been implemented heuristically, there can be a more precise algorithmic derivation of these lists that currently are manually created and managed for our case study.

String—one or more displayable ASCII characters delimited on the right side by a string delimiter (often a white-space). A string could include any other printing characters except delimiter(s), which are implementation specific.

Token—a string of alpha-numeric characters surrounded or delimited by whitespace.

Word—is a token where the sequence of alpha characters form a 'word' from a natural language and thus can be found in a lexicon. The same lexicon is used in this process to provide canonical source to consistently map words to TermIDs.

Lexicon—is a list of words that are pertinent to the domain of interest. Unlike a dictionary, a lexicon contains no other labeling or indications of semantics.

Synonym—a word that has nearly the same meaning (semantic) as another word in the corpus.

Synonym resolution—a process that identifies sets of words sharing a common semantic in the domain of interest. This set can be resolved to one representative word, and hence one TermID, prior to semantic analysis.

Named Entity—the proper- or given-name for an object. Named Entities thus refer to an object, but will not appear in a lexicon. When performing synonym resolution, named entities could be consistently replaced by a lexicon word of the class to which the entity belongs (domain specific).

Stop-word list—in a given natural language, there are certain word strings that are so frequently occurring and that have little semantic importance individually that to account for their occurrence frequency would dramatically skew the results. These words are enumerated in a stop-word list so that they may be removed from the documents universally and prior to tokenization to TermIDs.

Markup, container, and scripting stop-words—web traffic is communicated using the HTML markup language to present and render content within the various containers supported by the markup. Typical containers that utilize markup tags to convey structure and presentation are lists, text areas, tables, preformatted text areas, hyperlinks and targets, images, active scripting function code, and other non-textual content. Tag-names, certain attributes, and attribute-value pairs have no useful semantics and should be removed. While web-browser applications employ HTML and XML parsers to accurately separate content from markup, in typical traffic collection by a sensor network, we can not guarantee that web-pages and containers will be captured intact due to sampling techniques and the suppression of binary or non-ascii content. Therefore, we can not rely on a strict HTML or XML parser to properly differentiate all markup and scripting from content. JavaScript and other active content also present significant difficulties in processing prior to semantic analyses.

Auxiliary lexicon—one intent of our process is to recognize additional, non-English words that should participate in semantic analysis. This is commonly done in document classification of biology and chemistry corpora, where certain strings have high semantic content. Adding such strings to a lexicon will defeat the intention of a canonical representation of documents as TermIDs. We choose instead to form a more dynamic lexicon for each monitored IP address. During tokenization, we represent words by TermIDs from the canonical lexicon, and auxiliary strings by TermIDs from the address-specific lexicon. Each processing run will add new auxiliary words to this lexicon, and then assign the corresponding TermIDs from this lexicon.

The above definitions and concept extensions are important to allow the derivation and analysis of semantic and topic concepts from segmented, parsed, and transformed web-traffic. While these concepts and definitions are compatible with traditional semantic analysis of (English) prosaic documents, we anticipate that there are three significant differences in our process that allow more effective analyses of web-traffic:

1. The notion of stop word removal pertains as well to the removal of 'non-content' tokens. These are usually associated with HTML-code and HTML-format-markup.
2. Mixed alpha-numerics, and indeed pure or delimited numerics may also contribute significantly to semantic analysis for the purpose of web-traffic semantic characterization and comparisons. This is similar to the retention of numerics and non-word strings in bioinformatics.
3. Our notion of a web-document is formed by segmentation of surface-level (non-nested) GET or POST occurrences. In this manner, a document is formed by concatenation of all tokens observed from the initial request (GET/POST) through all subsequent responses, from possibly multiple servers and streams. With the natural, chronological order of traffic capture, this GET segmentation should faithfully represent all content related to the comparable GET so long as the session renders in just one browser-window (hence, reflecting the response of a single thread). The assumption we make is that all rendering in the one window pertains to the same set of topics as segmented by the GETs.

9 Processing Algorithms for Web-Traffic Segmentation and Tokenization

Our current process employs Perl scripts to prepare an ASCII traffic capture (file) to transform it into a document matrix file with (T:F) vectors per document.

TOKENIZE—reads a raw ASCII file and

1. Performs EOL/Newline removal to concatenate all lines into a pseudo-stream
2. Performs segmentation—by creating a new document each occurrence of 'GET' or 'POST' is observed.
3. The following punctuation is converted to white-space in cases where the punctuation is not acting as a delimiter:

```
$line =~ s/\x0A/*A*/g;        # remove newlines
$line =~ s/\x0D/*D*/g;
$line =~ s/\n/ /g;            # remove puncts
$line =~ s/[\.\,\;\*\(\)\'\"\`]/ /g;
$line =~ s/GET /GETT /g;      # prevent deletion
                              # of GET
```

4. Non-printing character glyphs, shown as period characters ('.') are inserted by the traffic capture rendering process. These are converted to spaces. Resulting white-space is collapsed into just one space, thus forming a GET/POST delimited set of tokens.
5. A token is emitted to the output file if it has a length > minTokenLength. The default is 4 characters. (Because of this, we replace 'GET' by 'GETT' to allow this token to be retained, along with the corresponding POST method-verb under default conditions).
6. This tokenization has preserved all potentially interesting tokens of non-punctuation printable characters of at least a minimum length.

DOC2MAT—performs lexical analysis, transformation, and output representation in (T:F) vectors, preserving word-oriented tokens and document segmentation. **DOC2MAT** is a re-implementation of the **doc2mat**.pl utility provided by the CLUTO project from George Karypis's lab at University of Minnesota, circa 2004. The original **doc2mat**.pl performs:

1. Conversion of document tokens to vectors of Term:Freq (T:F) pairs per document record
2. Optional: suppress standard stopword removal (-nostop)
3. Optional: remove all numerics or mixed alpha-numerics (-skipnumeric)
4. Optional: use custom stopword list (-mystoplist=file)
5. Optional: suppression of Porter Stemming (-nostem)
6. Optional: minimum word length (-minwlen=4), default is 3
7. Optional: emission of tokenized reassembly of each document to a tokenized file.

Our reimplementation of DOC2MAT is augmented to use a common (canonical) lexicon as a stable basis for TermID consistency across multiple batch processing runs. As such, each TermID always represents the same word, and we assume that each word asserts the same semantic when emitted from topic analysis. Some restrictions or complications include: domain specific semantics of the same word, synonym resolution in the domain, and named-entities which should also be resolved to an appropriate synonym. **DOC2MAT** performs the following:

1. The custom stop-list is first applied. After each execution run of **DOC2MAT**, a new-word list is written to a file for subsequent re-use. This new-word list will contain: non-words, named-entities, and non-content words like some markup and code tokens (words that often appear as valid words in the lexicon). It is most important that these non-content words be removed using this custom stop list. The current code and markup stoplist was manually created from a small set of typical web-browsing session documents. Although not complete nor totally accurate, it is likely to suffice for subsequent web-traffic pre-processing. A more formal construction would use the keywords from the JavaScript language and keywords from the HTML/CSS language specifications. It is likely that the removal of these words will not significantly impair the subsequent semantic/topic analysis.

2. Attempts to transform each token into the TermID from the lexicon.

3. If the token is not found in the lexicon, then it is added to a new-token list and assigned a new unique TermID.

4. Document (T:F) vectors are emitted, using the WORDID{$word} structure which consists of all words found in the corpus. TermIDs from the lexicon are always used consistently. New TermIDs generated by the retained, non-word tokens are appended to this combined lexicon and should be archived and associated with the specific IP address monitored in this corpus. TermIDs from new-words could later be added to a lexicon should we wish them to be mapped consistently across systems.

5. The first run for a corpus will generate (T:F) pairs corresponding to words in the lexicon and for new-tokens found in this corpus. By copying the new-word list to the custom stopword list, a second run will suppress all of these non-word tokens if that is the desired result. Alternatively, the new-word list can be edited, removing interesting words that should be retained in the output when the resulting subset is used as the new custom stoplist.

10 Lexicon Selection

The lexicon chosen for this work is the *2of12inf.txt* lexicon from Kevin Atkinson's site (http://wordlist.sourceforge.net). This lexicon was selected for this work due to the following features:

1. Familiarity (this author has used this lexicon frequently for applications in word-games and cryptography).

2. The desire to use a relatively large lexicon to optimize its utility for positive-filtration of web traffic. This is a primary means to select tokens that relate to topic content semantics rather than markup or code script tokens. This lexicon version from the year 2003 consists of 81520 words.
3. An overt preference for American-English words, including expected plurals, and hyphenations, again in an effort to recognize tokens related to content semantics.
4. The need to select an established and stable lexicon, since it is used to generate documents and corpora where each observed word is represented by the same TermID for all past and future documents.

11 Semantic Analysis

As we stated before, the focus of this work is the preparation process to transform collected web-traffic into a (T:F) representation form that could allow effective semantic analysis and document classification treatment. We feel it is important to at least show the effectiveness of our process by inspection of output from a typical analysis algorithm. We choose latent Dirichlet allocation (LDA) as representative of a modern, Bayesian-model approach for corpus analysis.

Since LDA is a parametric algorithm, we need to select appropriate parameters such as the anticipated number of topics (N) and also a rank-sensitive topic word selector function. In our small case study from collection of a student's web traffic for a typical day, we selected N = 10 topics, and for each such topic, selected the top or most significantly weighted 10 words per topic. These were selected ad-hoc during the refinement of processing and are certainly not optimal.

In future work, it may be desirable to retain larger topic word sets, providing their weights as well, so that threshold or mass-distribution interpretations can be applied for different types of analyses. Non-parametric analyses could also be used and likely combined with a mass-distribution sensitive interpretation.

12 Results

For our case study, we deployed a collection sensor on a small network, and collected roughly 24 h of web traffic between students and the internet, much of which came from the use of an academic, online learning portal. Tcpdump (actually Windump) was used to collect all traffic from a router span-port on the inside network. Parameters supplied to Windump caused it to collect all web traffic, retain complete, non-truncated packet contents, and write the collected traffic to separate files depending upon the IP address of the internal system-browser.

• Web traffic collected for the selected system amounted to 2.1 MB on the day of collection.

- WebStops list was 3 KB from 297 tokens of HTML, container, and JavaScript markup.
- OtherStops list was 23 KB from 2248 tokens, manually labeled as variable names and scripting attributes.
- 2of12inf.txt lexicon is 857 KB from 81520 English words, including plurals.
- Auxwords list was 45 KB from 5808 words representing named entities and significant identifiers, retained for this IP address.
- The TermID space was 87326 unique token-words, although some TermIDs for words in the intersection of the lexicon and the markup stopword list will never be included.
- After lexicon-based tokenization, the (T:F) document matrix retained 218 documents and was 157 KB.
- We compiled the LDA application in C from Daichi Mochihashi at NTT Communications Science Laboratories [8] on a dual-core Lenovo T60 with 2 GB RAM and running WinXP.
- Algorithms corresponding to the various components of the process were developed in object-oriented Perl(5) using both the Win32 and Linux64 environments. The only required Perl module was Getopt-Std for handling of command-line arguments.
- LDA processing converged in 12 s after 20 overall iterations.
- The output from LDA produced an 'alpha' vector and a 'beta' matrix of size $(87326 \times 10 \text{ FLOATS})$ corresponding to the contributions from the words represented in the rows, to the $N = 10$ columns of the beta matrix.
- The function 'unLDA' operated to sort each column in the beta matrix and presented the top 10 words representing each of the 10 topics.

Samples and excerpts from this case study appear in the Appendix.

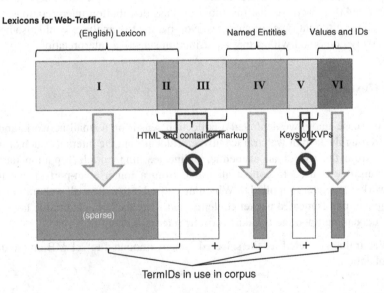

Fig. 1 Various lexicons used in our process to represent and analyze web-stream content. Three of the lexicons are (mostly) removed using stop-word lists. These sub-lexicons (II, III, V) would distract from semantic analyses that use T:F representations

13 Summary

We have developed methods for processing of web-traffic streams, preparing them in a manner that retains IP system specific lexicon-based representations for subsequent semantic analysis. Tokens and words that are deemed to have interesting contribution to semantic analysis are selected by a set of lexicons and stopword lists. In a small case study, the results of semantic analysis using LDA produced a set of 100 tokens for 10 topics. Cursory examination of these results show a roughly equal proportion of English lexicon words and numerical IDs or entities which seems to be a reasonable summary of the web-traffic collected. Future work should develop a more deterministic creation of the lexicons and stopwords, as well as investigate the opportunities afforded by non-parametric analyses of classifications and topics.

A.1 Appendix: Sample Representations

The following excerpt illustrates typical web-traffic captured by Snort, TCPdump, and other string-oriented capture tools. These tools often add line-feeds following header records to enhance human readability. Other, binary data are rendered as ASCII characters, or as '.' when the corresponding byte is not a printable character (Fig. 1).

A.2 Raw Data (from TCPdump)

```
17:51:47.430269 IP 209.133.xx.yyy.80 > 10.10.10.241.2798:
P13442:14899(1457) ack 3803 win 14060
E...Q%@....'..J...P.b))....OP.6.G\.. valign="top"><td width="50%"
valign="top" ><a href="#endNav_489037_1"><img src="/images/spacer.gif"
alt="Skip Module: Goodwin College Student Annoucements" height="1"
width="1" border="0"></a><table border="0" bgcolor="#000000"
cellspacing="0" cellpadding="1" width="100%"><tr><td><table border="0"
bgcolor="336699" cellspacing="0" cellpadding="2" width="100%"><tr><td
bgcolor="336699" width=5><img src="/images/spacer.gif" width=2
alt=""></td><td width="100%" bgcolor="336699" ><a name="Goodwin College
Student Annoucements"></a><h2 class="moduleTitle"><font color =
"FFFFFF">Goodwin College Student
Annoucements </font></h2></td></tr></table><table border="0"
cellspacing="0"
```

A.3 Raw Traffic Strings After Tokenization

```
GETT HTTP/1 Accept: Accept-Language: en-us User-Agent: Mozilla/4
compatible MSIE Windows
GETT /csi?v=3&s=webhp&action=&e=17259 28143 28506 28662 28832 28986
29013&ei=5dRqTdbkIsW3tgf18tz0DA&expi=17259
GETT HTTP/1 Accept: image/gif image/jpeg image/pjpeg image/pjpeg
application/x-shockwave-flash application/vnd
GETT /edgedesk/cgi-bin/login exe?bind=mail ascus HTTP/1 Accept: image/gif
image/jpeg image/pjpeg image/pjpeg
```

A.4 After Stop-Word Removal and Mapping to the Lexicon TermIDs

A validation display of token expansion from (T:F) back to lexicon word[T] per document. Documents #25 and #26 are shown:

```
[25] 33184:1 79943:1 406:1 2633:1 5958:2 59866:1 58600:1 53364:1 79078:1
28355:1
[25] here which access another being replace redirected please wait
forwarded

[26] 72729:1 31229:2 17093:1 62691:1 39238:1 40394:1 40657:1 21162:2
1979:1 1926:1 12983:1 35072:2 56748:1 51620:1 23929:7 48837:1 29048:1
77649:1 61498:1 3391:3 45458:3 73217:5 3614:2 75614:1 77659:1 1924:1
80282:2 46667:1 77825:2 79908:2 25143:1 406:3 80569:2 79279:1 46662:2
[26] these greeting customized save keywords least less docs always also
code illegal qualified pending entered open functions used role area month
time artifact unable users already will navigational validate when
```

A.5 WebStops

The webstops list contains many word tokens that comprise HTML markup and containers in web pages, such as tables, JavaScript functions, list structures, and style sheets. These words relate to the construction of such containers common to all web pages, and hence, are devoid of semantic content. These tokens, even though colliding with some lexicon words, must be removed so that the high frequency of these tokens does not dominate (T:F) sensitive semantic analysis algorithms like latent Dirichlet allocation.

academic	block	chars
accent	body	check
accept	bold	class
action	border	click
agent	bottom	clip

(continued)

(continued)

align	bounding	close
alive	box	color
application	boxes	colorful
author	browse	comma
auto	browser	common
background	bundle	compatibility
banner	button	compatible
batch	buttons	connection
before	bytes	console
begin	cache	content
bind	cancel	continue
blackboard	center	control
blank	char	cookie
	character	cookies
	characters	copy
		. . . .

A.6 CustomStopWords

Modern web pages contain other non-lexicon words associated with JavaScript code, variable names and values. JavaScripting markup attributes also typically contain numerous key-value pairs. We observe that the values of KVPs most often contain semantically interesting content. The current process tries to retain these values, as well as other named entities, while removing non-lexicon keys and variable names. The following custom stopword list was formed after examination of a small set of web-pages, and was labeled by the author for subsequent use on this dataset. We also suggest that a more accurate and dynamic result should process the KVPs early in tokenization by splitting on the equals character ("="). HTML and JavaScript keywords can be formally enumerated and removed. Variable names will be more difficult to determine precisely; it is likely partial-word stemming of segments may yield satisfactory performance since most often, variables consist of concatenated words, sometimes camel-cased, for self documentation. Entropy measures of the discovered components could also improve recognition of variable names.

bbnj	puvq	carin
pbtpid	panose	callout
errorh	btngradientopacity	imcspan
yvlq	abpay	unexpectedtype
validatedelete	pubi	headerbgcolor
logout	rssheadlinecell	colheader
classe	brea	codebase

(continued)

(continued)

iptg	serv	privacypolicy
sfri	offborder	jrskl
emihidden	regexpmatch	pollcometinterval
pickname	fieldcaption	reqrevision
headgrade	playlists	baccentmedium
mathfont	getmenubyname	substr
nbsp	clickimage	bord
sethttpmethod	nprmodpipe	active
		. . . .

A.7 Semantic Analysis After LDA (Latent Dirichlet Allocation)

Treating such vectors in their pre-converted form would preserve anonymity while allowing trending, differentiation and anomaly analysis and comparisons. The following output has been expanded to the original words as a validation step to show correspondence to the original corpus documents.

```
TOPIC: 0 --------------------------------
  17520 2011 wjql 306627 30729 4506 2152 50727 1460 course
TOPIC: 1 --------------------------------
  1460 6432 17520 2864 2861 2803 2011 7504 version 1466
TOPIC: 2 --------------------------------
  1460 17520 2842 2832 2813 prop 7544 batches call from
TOPIC: 3 --------------------------------
  course 1460 time 17520 252320 assessment 284540 2011 6432 alert
TOPIC: 4 --------------------------------
  1460 entered than 17520 deployment course contain announcement i18n time
TOPIC: 5 --------------------------------
  calendar discussion forum board entry course 1460 month 2811 284540
TOPIC: 6 --------------------------------
  calendar month 1460 discussion course forum board entry 2830 long
TOPIC: 7 --------------------------------
  forum your 2815 course blight 284540 discussion 1460 board 3774
TOPIC: 8 --------------------------------
  calendar course forum discussion board 1460 2835 2831 your month
TOPIC: 9 --------------------------------
  2851 65392 12px 2852 2844 16616 solid repeat hover 1460
```

References

1. C. Wang, D. Blei, D. Heckerman, *Continuous Time Dynamic Topic Models* (Princeton University, Princeton, 2008)
2. M. Hearst, *Multi-Paragraph Segmentation of Expository Text* (Computer Science Division, UC Berkeley, Berkeley, 1994)
3. A. Jain, A. Kadav, J. Kawale, Semantic Text Segmentation and Sub-topic Extraction. Retrieved from http://citeseerx.ist.psu.edu/viewdoc/download?doi=10.1.1.120.7624&rep=rep1&type=pdf, 2008

4. R. Kern, M. Granitzer, Efficient linear text segmentation based on information retrieval techniques. *MEDES 2009*, Lyon, France, pp. 167–171, 2009
5. M. Porter, An algorithm for suffix stripping. Program **14**, 130–137 (1980)
6. R. Futrelle, A. Grimes, M. Shao, Extracting structure from HTML documents for language visualization and analysis. Biological Knowledge Laboratory, College of Computer and Information Science, Northeastern University, in *ICDAR (Intl. Conf. Document Analysis and Recognition)*, Edinburgh, 2003
7. P. Wittek, S. Daranyi, Spectral composition of semantic spaces, in *Proceedings of QI-11, 5th International Quantum Interaction Symposium*, Aberdeen, UK, 2011
8. D. Mochihashi, lda, a Latent Dirichlet Allocation package. NTT Communication Science Laboratories, 2004. http://chasen.org/~daiti-m/dist/lda/
9. G. Stumme, A. Hotho, B. Berendt, *Semantic Web Mining State of the Art and Future Directions* (University of Kassel, Kassel, 2004)
10. J. Williams, S. Herrero, C. Leonardi, S. Chan, A. Sanchez, Z. Aung, Large in-memory cyber-physical security-related analytics via scalable coherent shared memory architectures. *2011 IEEE Symposium on Computational Intelligence in Cyber Security (CICS)*, 2011
11. P. Wittek, S. Daranyi, Connecting the dots: mass, energy, word meaning, and particle-wave duality, in *QI-12, 6th International Quantum Interaction Symposium*, Paris, France, 2012

Concurrent Learning Algorithm and the Importance Map

M. R. McLean

This chapter describes machine learning and visualization algorithms developed by the Center for Exceptional Computing, a Department of Defense research laboratory. The author hopes that these tools will advance not only cyberspace defense related applications, but also a number of other applications where cognitive information processing can be integrated. The chapter begins by describing the difference between conventional and cognitive information processing. There are pros and cons for using either of these approaches for problem solving and the ultimate decision of which to use is entirely related to the problem at hand. However, this chapter will focus on the cognitive approach and introduce the algorithms that were developed to make the approach more attractive. The Concurrent Learning Algorithm (CLA) is a biologically inspired algorithm, and will require a brief introduction to neuroscience. This introduction is necessary to familiarize the reader with the unique biological aspects that form the foundation for the CLA; the detailed algorithmic description will follow. Finally, the Importance Map (IMAP) algorithm will be introduced and examples given to clearly illustrate its benefits, spanning from neural network development to end user applications. By the end of this chapter the reader should have a firm understanding of the unique abilities afforded by these algorithms and be able to implement them into code.

For the past 60+ years information processors have been typically based on the Harvard or Von Neumann architecture. These processors require four steps to process one instruction: (1) fetch the instruction and data from memory; (2) decode the instruction; (3) execute the instruction on the data; and (4) store the result to memory. Each processed instruction does one basic operation and a program can easily consist of millions of instructions. Using this paradigm is referred to as an algorithmic approach. One obvious issue with this approach is that it restricts the problem solving capability to algorithms that are precisely defined. This restriction

M. R. McLean (✉)
Center for Exceptional Computing, Catonsville, MD, USA
e-mail: mrmclea@lps.umd.edu

R. E. Pino (ed.), *Network Science and Cybersecurity*,
Advances in Information Security 55, DOI: 10.1007/978-1-4614-7597-2_15,
© Springer Science+Business Media New York 2014

sounds reasonable, however, there are many problems that do not have an exact algorithm or that change over time. Another problem with the algorithmic approach is that it requires specialized skills to translate algorithms into programs and often requires that the system be developed by different individuals. The software development process can be slow, expensive, and prone to errors. Furthermore, the developed software executes a very specialized, rigid program that cannot adapt or improve over time. This is where neural networks may provide a better solution than the traditional software development method.

Neural networks use a non-algorithmic approach to problem solving. They are trained by example, similar to the human learning process. The neural network is presented data and asked to evaluate it; if it is evaluated incorrectly the network weights are adjusted. Training a neural network does not require learning a special language, just examples with correct classifications. Neural networks have a remarkable ability to derive understanding from highly dimensional, complex, and imprecise data. They are able to extract patterns and detect trends that are too complex to be noticed by either humans or other computer techniques. They learn without bias and can discover relationships that are not intuitively obvious. Over time a neural network becomes more accurate as it is presented more data, eventually becoming an "expert" at the selected concept.

So why aren't more people using neural networks? Neural networks are not perfect, and their same strengths can also be their weaknesses. There are certainly issues to implementing a reliable neural network system. As mentioned, neural networks are unbiased, finding any correlation of the inputs to the outputs. This is exacerbated as the dimensionality of the data increases, a real world example should clearly illustrate this problem. A neural network system was created to differentiate between enemy and allied tanks. Many images of the both enemy and allied tanks were taken and used to train the neural network. The neural network achieved very high accuracy in the lab, so a field demonstration was scheduled to show the program manager the system capability. On the day of the demonstration the neural network performed miserably. A lengthy analysis showed that the pictures of the enemy tanks were taken on a sunny day and the pictures of the allied tanks were taken on an overcast day—the neural network had classified the tanks based on sunny or overcast days. Until recently, the only way to know if a neural network had learned the intended concept correctly was to run many training and test iterations with as much training data as possible. The example above shows this approach does not scale well with increasing dimensionality and has resulted in many other failed implementations. Neural networks have always been a fairly opaque black box with minimal visibility into what was actually learned, impeding the adoption of neural networks into the mainstream. Another problem with neural networks is they classify data without giving any insight into the reasoning behind the classification. This makes it difficult for an end user to trust the system and often leads to a failed implementation. Later in this chapter the IMAP algorithm will be discussed, it shows the concepts the neural network has learned and resolves the many of problems discussed above.

The CLA is called as a supervised learning algorithm, meaning the neural network requires correctly classified data for training. The dataset contains a series of input vectors and the associated target answer. CLA uses a process called gradient descent to minimize the network error; however, only the direction of the gradient is used to modify the decision surface created by the neural network. The direction of learning is a delta value, computed by subtracting the target answer from the actual neural network output. Once the delta values for all the outputs have been computed, all nodes in the neural network can learn concurrently, hence the name of the algorithm. This unique ability greatly speeds up network learning and can take advantage of today's highly parallel architectures.

The basic element of the CLA neural network is called a node, which is similar to a neuron in the cerebral cortex of the brain. A node has a number of inputs connected to it called edges, which are similar to the dendrites of a neuron. The node function of CLA is very similar to other supervised learning algorithms except there is an added attribute to each of the edges.

Figure 1 is the diagram of the CLA node. The input vector is designated $x_1 ... x_n$ and each of it's elements connects, via a graph edge, to the node. The added attribute unique to CLA is the influence value denoted by $S_{j,i}$ and the final edge attribute $W_{j,i}$, represents the synaptic weight that gets adjusted during training. Every feed forward connection will have these three attributes. The influence is unique to CLA and its origin and importance are discussed next.

In the brain there are billions of neurons and each occupy a unique location in 3-D space. Every neuron has an output connection called an axon. The axon conducts a pulse called an action potential that propagates information to other neurons. There are two basic types of axons in the brain: myelinated and unmyelinated. The myelinated axons are used to connect the hemispheres in the brain and control muscle movements. The myelin sheath covering the axon results in the action potential having low latency and low resistance, enabling long distance neural communication. Without myelinated neurons, multi-sensory models in our brain would be difficult or impossible to develop. The myelinated axons have a white appearance and compose the white matter in the brain. However, the bulk of the computation in the cerebral cortex takes place using unmyelinated or grey matter axons. Being unmyelinated, the activation potential or signal of these axons degrade in strength and speed over their lengths. This limited interaction creates "a sphere of influence"—as the length of the axon increases its influence on other nodes decreases according to the inverse square law.

Fig. 1 CLA node

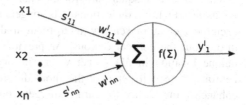

Fig. 2 Influence versus
distance

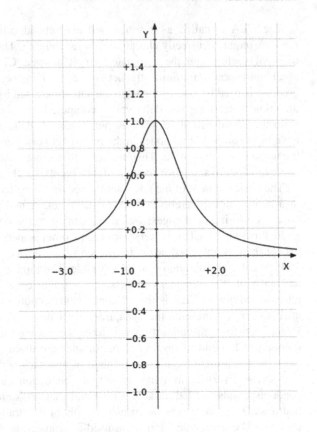

Figure 2 shows how the influence between nodes decreases over distance; the horizontal axis is distance between nodes and the vertical axis is influence. A simple non-temporal explanation is to consider the termination connections of the axon as a cone, as the connection length increases and the cross sectional area decreases quadratically. The electronics equation of the resistance of a conductor clearly supports this hypothesis.

$$\Omega = \frac{\rho \times L}{A} \qquad (1)$$

Equation 1 is the equation for the resistance (Ω) of a conductor where ρ is the conductance of the material, L is the length of the conductor, and A is the cross sectional area in units^2. Since ρ is a constant, it can be ignore in the equation, leaving an inverse square law for the change in resistance as the length increases. As its termination connections gets longer, an axon's ability to influence nodes decreases, creating sensory interpretation diversity and functional partitioning throughout the cortical regions. As previously mentioned, all neurons occupy a unique location in 3D space; when sensory information is presented to these neurons, each will perceive the input differently, which increases diversity and enhancing the ability to learn. Emergent functional partitioning, created by the influence, enables many local computations to occur simultaneously and independently.

CLA integrates this influence into its learning mechanism. The influence attribute for the connections of the network is calculated by first setting up the node layout and layer mappings in 3D space. In CLA, the node layout and layer mapping algorithms are completely flexible and can be adjusted to best suit the problem. In the cerebral cortex the nodes are laid out in six distinguishable striations or layers. Figure 3 shows how CLA uses a similar node layout; in this example there are only four layers, however CLA has no restriction on the number of layers in the network. The four layers in Fig. 3 are, from bottom to top: the input layer, hidden layer, output layer, and a delta layer. Each of these layers will be discussed shortly. For now, note that the nodes position in 3D space will be important for influence computation.

After laying out the nodes, CLA connects nodes between the layers in a process called layer mapping. Figure 4 shows a partial mapping between layer 0 and layer 1 due to influence thresholding. This means only nodes that meet or exceed a certain threshold influence are connected to each other. The traditional supervised learning approach is a fully connected network, where the current layer connects to all nodes on the previous layer. Fully connected neural networks are computationally intensive and more susceptible to noise. For example, consider a fully connected network with a lower layer of 10,000 nodes—every node in the current layer would have 10,000 connections. If there is a significant signal on only one of the node's inputs, the signal-to-noise ratio (SNR) would be 1/10,000, or 0.00001. By partially connecting nodes based on their influence, the SNR of the network can be adjusted to better suit the problem. In this example, restricting the current nodes to connect to only the 100 most influential nodes in the previous layer would increase the SNR by two orders of magnitude, thereby increasing the overall

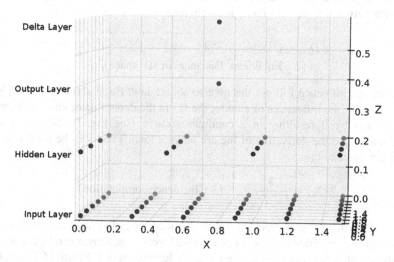

Fig. 3 CLA node layout

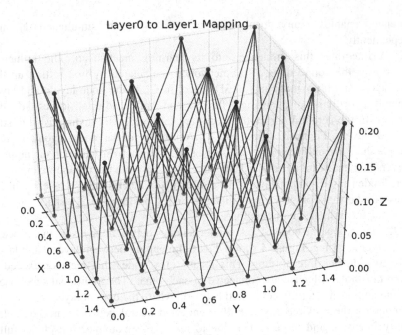

Fig. 4 CLA partial connectivity

network stability. One drawback to this approach is that it may require additional layers to be added to the network.

The CLA nodes use influence during computation and to limit connectivity between layers. The first step in calculating the influence is to compute the distance from one node to another node in 3D space, accomplished using the Euclidean distance formula for three dimensions (Eq. 2). In this equation The distance between node$_j$ and node$_i$ is calculated.

$$D_{j,i} = \sqrt{(x_i - x_j)^2 + (y_i - y_j)^2 + (z_i - z_j)^2}$$
(2 : Euclidean Distance in 3D space)

Since the influence follows the inverse square law, Eq. 3 shows the complete calculation of the influences of a node; the '1' in the denominator ensures there is no division by 0 resulting in a continuous curve (see Fig. 2). Note that this equation is also the derivative of the arc tangent (atan') and will be used later in the CLA calculations.

$$S_{j,i} = \frac{1}{1 + D_{j,i}^2} \qquad (3 : \text{Influence computation})$$

The CLA mapping is repeated for all connections in the network. CLA is not restricted to connections between consecutive layers—each layer can be connected to any and all following layers, increasing the learning speed. Finally, CLA makes use of feedback connections between the top most layer of nodes (the delta layer) and the lower layers. The feedback mapping is as flexible as the feed forward connection; however, every node besides an input nodes has to be connected to at least 1 delta node. The feedback connection from the delta nodes have an input and an influence, but no weight. The input to the feedback connection is the output of the delta node, which, as discussed previously, is target minus actual output. This delta value gets scaled by the influence of the connection and is used by the node for learning.

Once the influences for all the network connections are calculated, there is no further need to consider the neural network in 3D, and for clarity normal 2D diagrams will suffice. As mentioned, Fig. 1 shows the model of a single node in the CLA algorithm; these nodes are connected together to create a neural network (Fig. 5).

This network has an input layer (on the left) with three solid nodes, one hidden layer (in the middle) with three nodes and an output layer (on the right) with three nodes. The outputs actually feed into the delta nodes described earlier; however, the delta layer is only used during training and is responsible for sending difference between the target and actual signals to each node. There is an additional input to all hidden and output nodes called the **bias**, which sets thresholding for node firing. The input and influence of the **bias** is always 1, so the connection weight is simply added to the sum of the node.

Supervised learning networks operate in two modes: training and evaluation. In training mode, a randomized percentage of the data, normally 80–90 %, is presented to the neural network. Weights are adjusted to minimize the overall error between the target data and the actual output of the neural network. When the error has been sufficiently minimized, the other 10–20 % of data is used to test how well the network has learned the concept. This accuracy of the neural network on the test dataset is a measure of how well the neural network generalizes. Cross validation can also be used to better characterize the network's generalization capability. For example, completing only a single test without any cross validation has a possibility that the dataset was created with test vectors nearly identical to the ones used to train the network. This would be a very biased test case and is the

Fig. 5 Network structure

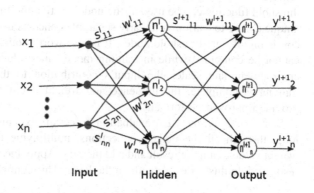

reason that some method of cross validation is normally used. Cross validation trains the network multiple times, holding back a different test set for each session and provides assurance that the network has learned correctly. It also gives an indication that the size of the dataset is sufficient for the concept being learned.

Both the training and the test datasets contain an input vector and a target vector. The network completes an *epoch* when all the vectors in the training dataset are presented to it. Normally, the order of the training vectors are then randomized and another epoch is completed. This process repeats until the desired error between the target and actual output of the network meets user defined requirements. There are two methods of training: *on-line* and *batch* mode. In *on-line* training, the weights are adjusted after each input vector is applied. *On-line* training could potentially be very fast, however, if the dataset is very large, and the concept complex, then the neural network may have problems learning. Batch mode only updates the weights only at the end of an epoch. While very slow, this method better portrays the overall gradient of the error surface. CLA uses both of these learning methods; the one applied is based on the characteristics of the problem and the dataset.

For the *on-line* method, consider the input vector represented by $\mathbf{X}$ and containing $\mathbf{n}$ number of elements. These inputs are real valued numbers and a dataset would contain multiple input vectors resulting in a 2D array or matrix. Every input vector in the dataset must contain the same number of elements. When the input vector is applied, the network evaluation proceeds layer by layer starting with the first hidden layer and ending at the output layer. Equation 4 shows the formula for CLA node evaluation (for further clarification refer to Fig. 1).

$$DP_j = \left(\sum_{i=0}^{n} (x_{j,i})(s_{j,i})(w_{j,i}) \right) + w_{bias} \qquad (4: \text{CLA node evaluation})$$

$$f(DP_j) = Threshold(x), atan(x) \qquad (5: \text{Activation function})$$

In Eq. 4, $\mathbf{x}$ is the ith input to the jth node, $\mathbf{s}$ is the influence of the connection, and $\mathbf{w}$ is the synaptic weight. The CLA node computes a dot product or multiply accumulate operation on all of its inputs, adds the bias, then applies the activation function to the dot product to determine the actual output. Equation 5 shows that a threshold function can be used as the node's activation function. CLA also has the unique ability to learn non-linearly separable concepts using a multi-layer network consisting of only threshold nodes. It is also worth noting that the $\mathbf{s}$ and $\mathbf{w}$ values cannot be combined while in training mode—the influence reflects the system's environment and scales the input's contribution to the node. However, after training is complete, the influence and weight could be combined and would provide improved performance.

Systematically, the CLA evaluation process is explained as follows: for node j loop, through all the nodes connections, multiplying the input, influence, and weight values, sum the result and add the bias. Apply the activation function to this result and do this for every node in the layer. This continues for every layer in the network until the output layer is reached, meaning the network has been evaluated and the **actual** values are Y_n^{output}. Using the **actual** outputs and the **targets** values from the training data, the node's direction of learning can be computed. The formula for the direction calculation is shown in Eq. 6.

$$Dir_j^l = \frac{\sum_{i=0}^{outputs}(Target_i - Actual_i)S_{j,i}}{outputs} \qquad (6 : \text{Learning direction})$$

This equation calculates the mean learning direction $\left(\mathbf{Dir_j^l}\right)$ of the jth node on the lth layer. The delta value (**Target-Actual**) for the ith output is scaled by the influence **S** of the ith delta node connection. This is done for all the outputs connected to the node, each result is summed together and the total sum is then divided by the number of outputs. The mean learning direction is *not* dependent on intermediate calculations from consecutive layers like back propagation. This implies that given enough hardware resources, all the nodes in the neural network could be trained concurrently.

The next value computed is the derivative of the atan'(**DP**), which is similar to the back propagation computation of partial derivatives using the chain rule. In Eq. 7 **DP** is the dot product as calculated in Eq. 4. The computation contributes to a unique learning rate for each node in the network and also bounds the synaptic weight values. As the dot product of the node increases, the amount of learning is reduced non-linearly similar to the influence (see Fig. 2).

$$Atan'(DP_{j,i}) = \frac{1}{1 + (DP_{j,i}^2)} \qquad (7 : \text{Arc tangent derivative})$$

The arc tangent derivative is the last calculation needed to complete the weight update. For each node in the neural network a delta change amount is calculated as shown in Eq. 8. This is the total amount of change the node will make and is then scaled by each connection in proportion to its impact (see Eq. 9).

$$\Delta_j = \eta\frac{Dir_j}{(1 + DP_j^2)} \qquad (8 : \text{Node change computation})$$

$$w_{j,i} = w_{j,i} + \Delta_j(x_{j,i})(s_{j,i}) \qquad (9 : \text{Weight update})$$

In Eq. 9, $\mathbf{w_{j,i}}$ is the weight of the ith input, η is a constant learning rate of the network, $\mathbf{x_{j,i}}$ is the ith input of the node and $\mathbf{s_{j,i}}$ is the associated influence of the connection. The calculation is repeated for all connections to the node, then for all the nodes in the network and completes the weight modification for the error produced by the input vector. This process continues for each input vector in the training set and the accumulated mean squared error (**MSE**) is returned at the end of each epoch. Equation 10 shows the MSE calculation, where **N** is the number of input vectors in the dataset, **O** is the number of outputs, **target** is the expected output value, and **actual** is the network output.

$$MSE = \frac{\sqrt{\sum_{i=1}^{N} \sum_{j=1}^{O} (target_{i,j} - actual_{i,j})^2}}{NO} \qquad (10 : \text{Mean squared error})$$

The neural network is trained epoch by epoch until the MSE is reduced to a user defined value. The network is then tested to ensure it generalizes correctly. The test vectors are presented consecutively and the neural network is evaluated; any errors recorded are presented to the user. Here some form cross validation would ensure the generalization capability of the network, but that requires the neural network to be trained and tested repeatedly by selecting different training and test vectors from the dataset.

Although CLA brings many benefits to supervised learning, it still has the discussed weakness of being a black box. Neither designer nor end user understands the reason for the classification. Even with the testing and cross validation methods, CLA still faces the *'curse of dimensionality'*: when the input vector has a high number of elements and the number of input vectors is proportionately small, there is a good chance there will be coincidental correlations that have nothing to do with the desired concept. The IMAP visualization addresses this issue by showing how each element of an input vector contributes to a particular output value (classification). If the neural network had five outputs, there would be five unique importance maps created for the applied input vector. The importance maps created are the same size as the input vector, and can be represented in 1, 2 or even 3 dimensions. To illustrate the usefulness of this tool, consider an example using the face images data set downloaded from the University of California Irvine's machine learning repository. There are 640 greyscale images each at a resolution of 120×113 pixels, for a total input vector length of 13,560. The image values were normalized from -1 to 1 and the network was trained to identify people wearing sunglasses. The CLA achieved $\sim 96\ \%$ accuracy against this dataset.

In Fig. 6, the green overlayed dots on the screen indicates those pixels were important to be dark and the red indicates they were important to be light for the classification. Clearly, the concept of wearing sunglasses has not been learned completely, since there are many important elements not on the facial region. While the network achieves a $\sim 96\ \%$ accuracy on test cases and may indicate the concept has been learned, the importance map shows considerably more information. Figure 7 shows the IMAPs for the a needle in a haystack (NIH) dataset, a synthetic dataset where a single pixel is correlated to the output. The NIH data represents a set of 16×16 8-bit greyscale images.

Figure 7 shows how the IMAP can take a highly dimensional dataset and display the single element responsible for the output. Also, in both previous cases the IMAP can be used to reduce the dimensionality of the dataset. Eliminating unneeded input vector elements directly reduces the noise and leads to a more stable neural network. Dimensionality reduction also decreases the amount of computation required to train and evaluate the neural network. A single IMAP alone should not be used to reduce dimensionality, since the IMAP shows the important input elements for a particular input vector. A more sound approach

Fig. 6 IMAPS of sunglasses recognition

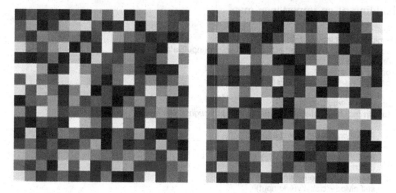

Fig. 7 NIH IMAPs

would be to average all the correctly classified vectors for a single output state (for NIH all vectors correctly classified as a 1). The average IMAP for this classification shows a more general concept of what the neural network considers important. Using the reduced input vectors may cause overall accuracy of the neural network to decrease, but the errors that occur may help identify incorrectly labeled input vectors or outliers. For all of these reasons, IMAPs prove to be a very useful diagnostic tool with profound impacts on fielded systems.

IMAPs are also be extremely useful for the users of the neural network. Normally, a neural network system delivers classification without explanation. In an anomaly detection system, the analyst only receives a classification of anomalous behavior and would need to look through the highly dimensional data to verify it is truly anomalous. If the network is wrong, the user has no idea what caused the error. This can be frustrating and lead to the user having a lack of confidence in the system. However, IMAPs would greatly reduce the analyst's search space, showing the user exactly why it classify the data as anomalous. Even if the classification is wrong, the user understands why and can see how the network improves over time—the network interacts with the user. This interaction can lead to increased customer confidence in the system and a willingness to continue to improve the system.

There are two different methods to build IMAPs; top-down and bottom-up. Both end up with the same result, but depending on your network configuration one may perform better. The top-down implementation uses a recursive algorithm that propagates from the output all the way down to the inputs computing each input elements importance. The bottom-up approach is a little more straightforward and will be discussed in detail. The first thing that needs to be done is store the dot product of each of the output nodes, then set an element in the input vector to 0, re-evaluate the network and subtract new output from the original. This difference is the impact that particular element had on each of the outputs. Once this is accomplished for every element in the input vector, the impacts are then normalized for each output IMAP. Below is a snippet of Python code to implement the building of an importance map.

```
1  def genImap2(self,inputData):
2      self.imaps = np.zeros((self.eModel.outLen, len(inputData)))
2      origInput=0.0
4      origDPs = list()
5      origDPs = self.getOutputDPs()
6
7      origInput = inputData[:]
8      for i in range(self.eModel.inputLen): #loop through all the inputs
9
10         #store original input value
11         bitSave = origInput[i]
12         #set the value to zero
13         origInput[i] = 0
14         #re-evaluate the network
15         self.eModel.classify(origInput)
16         #store the changes
17         newDPs = self.getOutputDPs()
18         for j in range(self.eModel.outLen): #loop through all the outputs
19             #check to see if the DP got smaller this means the input element  20 #helped the output value
21             val = abs(origDPs[j]) - abs(newDPs[j])
22             if val > 0.0: #this input contributed to the output response
24                 self.imaps[j][i] = val #store the element effect
25                 #if the input was negative multiply by 0
26                 if bitSave < 0.0:
27                     self.imaps[j][i] *= -1.0 #this will be green when
28                                                                                    displayed
29
30         origInput[i] = bitSave
```

This chapter discussed the difference between the traditional algorithmic approach to problem solving and the neural networks approach. It discussed the biological grounding for adding an influence on each of the neural networks connections and discussed the resulting emergent behavior. The chapter then explained some general principles of how to use neural networks, as well as the specifics of implementing the CLA. Finally, the IMAP algorithm was discussed and explained how this visualization algorithm is very useful to both the neural network designer and the end user. It was then discussed how the IMAP can be used to mitigate the curse of highly dimension data and give insights on concept the neural network has learned. With this information you should be able to harness the enhanced capability of these algorithms and apply them to your applications.

Hardware Accelerated Mining of Domain Knowledge

Tanvir Atahary, Scott Douglass and Tarek M. Taha

1 Introduction

Computer aided decision making is becoming increasingly important in a variety of applications. Two particular ones are in making sense of large amounts of data and in decision making in autonomous systems. Although humans are very good at decision making, computer aides are highly useful when dealing with large amounts of data and when high speed decision making is needed.

This chapter introduces a cognitively enhanced complex event processing (CECEP) architecture being developed at the Air Force Research Laboratory (AFRL) and examines the acceleration of this architecture to enable high speed decision making. The second section of the chapter describes the research effort underlying the development of the CECEP architecture. This description of the AFRL research effort contextualizes the research discussed later in the chapter and clarifies the importance of constraint satisfaction problem (CSP) acceleration within the architecture. The third section of the chapter formally defines CSPs [1–4] and illustrates the role the solution of CSPs play in the architecture. Example cognitive domain ontologies (CDOs), or formal representations of structural and relational domain knowledge, will show how CSP enables: (1) activity monitoring; and (2) inference from evidence to likely causes. Activity monitoring is a valuable capability in CECEP models and agents that are required to identify signatures of

T. Atahary (✉) · T. M. Taha
Department of Electrical and Computer Engineering, University of Dayton,
Dayton, OH 45469, USA
e-mail: ataharyt1@udayton.edu

T. M. Taha
e-mail: tarek.taha@udayton.edu

S. Douglass
Air Force Research Laboratory, 711 Human Performance Wing, Wright-Patterson AFB,
OH 45433-7955, USA
e-mail: scott.douglass@wpafb.af.mil

R. E. Pino (ed.), *Network Science and Cybersecurity*,
Advances in Information Security 55, DOI: 10.1007/978-1-4614-7597-2_16,
© Springer Science+Business Media New York 2014

intent. Inference from evidence to likely causes, a type of inference referred to as abduction, is a valuable capability in models and agents that are required to gather and make sense of observational data. Both capabilities can be exploited in cyber/physical system intrusion detection. The fourth and fifth sections of the paper examine related research efforts and examine the details of GPGPU architectures utilized. The sixth section describes research efforts to accelerate CSP in the CECEP architecture. This section explains how algorithms responsible for CSP in the CECEP architecture have been parallelized and evaluated on multi-core computers and GPGPUs. The seventh section provides results and analysis of the acceleration efforts while the last section concludes with a summary of findings.

2 Large-Scale Cognitive Modeling in AFRL

For the last 3 years, an AFRL Large-Scale Cognitive Modeling (LSCM) research effort has worked to close capability gaps retarding the development and fielding of advanced training, automated sense making, and decision support capabilities built using cognitive models and agents. Generally, the LSCM initiative has endeavored to enhance and accelerate the practice of cognitive modeling in AFRL. Specifically, the LSCM initiative has researched and developed:

(a) Domain-specific languages (DSLs) tailored to the needs of cognitive modelers.
(b) Authoring environments in which users employ these DSLs to specify models and agents.
(c) Code-generation technologies that transform models and agents specified in these authoring environments into executable artifacts.
(d) A cognitively enhanced complex event processing (CECEP) architecture in which models and agents are executed.
(e) A net-centric associative memory application [5] enabling models and agents executing in the CECEP architecture to store and remember declarative knowledge.
(f) A constraint-based knowledge representation and mining application [6] that allows agents executing in the CECEP architecture to produce and recognize actions using domain knowledge rather than fixed rules.
(g) Analysis and visualization capabilities that help cognitive scientists understand and analyze models functioning more like autonomous agents than programs.

To reduce model and agent development costs in AFRL research efforts, the LSCM research initiative is developing a DSL called the research modeling language (RML). The RML DSL is used by AFRL cognitive scientists to specify models and agents that execute in a net-centric CECEP architecture. The RML has been developed using the Generic Modeling Environment (GME); a meta-modeling tool for creating and refining domain-specific modeling languages and program synthesis environments [7–9].

In order to maximize scalability and interoperability during execution/simulation, RML and the GME-based authoring environment require users to conceive of and specify their models and agents as *complex event processing agents*. The graphical/textual RML DSL allows cognitive modelers to efficiently specify models and agents in this way using representations of:

Events: objects that serve as records of activities in a system. Objects capture the details of events with attributes and data properties. Event objects can be used to represent data that a decision aid is making sense of through inference and hypothesis generation.

Event Patterns: templates matching one or more events in an event cloud constituting a representation of context. Event patterns can be used to enable a decision aid to detect correlational, temporal, and causal relationships between events representing a decision situation.

Event Pattern Rules: associations specifying actions that occur after an event pattern is matched with a subset of events in an event cloud. Event pattern rules can be used to enable a decision aid to produce abstract events that combine and aggregate attributes of event patterns.

Behavior Models: sets of event pattern rules arranged into finite state machines that explicitly represent behavior-specific combinations of cognitive state, contextual factors, alternative courses of action, and failure. Behavior models can be used to enable a decision aid to monitor simple, complex, concurrent, and hierarchically organized activities and therefore make sense a complex decision situation.

Cognitive Domain Ontologies: representations of domain knowledge capturing: (1) entities, structures, and hierarchies in a domain; and (2) relations between these entities. Cognitive domain ontologies (CDOs) can be used to represent constraint knowledge a decision aid can process in order to develop and refine hypotheses relating observational data to likely causes.

Modeling in RML is based on the specification of procedural knowledge (behavior models) and domain knowledge (CDOs). Simulation of RML models and agents is based on the execution of code artifacts produced with code generators in the CECEP architecture. The CECEP architecture consists of the following central net-centric components:

soaDM: an associative memory application that allows RML models and agents to store and retrieve *declarative knowledge*. Declarative knowledge is represented and processed in a semantic network [5].

soaCDO: a knowledge representation and mining application that allows RML models and agents to store and exploit *domain knowledge*. Domain knowledge is represented in CDOs and processed by a constraint-satisfaction framework [6].

Esper: a complex event processing framework that allows RML models and agents to base actions on context assessment and *procedural knowledge*. Procedural knowledge is represented in RML behavior models and processed using pattern matching and event abstraction capabilities provided by Esper [10].

Through these components, the CECEP architecture incorporates model and agent capabilities based on declarative, procedural, and domain knowledge processing into the Esper framework. The resulting event-driven architecture is an advanced cognitive modeling and simulation framework with which AFRL cognitive scientists can develop and field decision aids and instructional technologies based on cognitive models and agents. A functional representation of the CECEP architecture is shown in Fig. 1.

The CECEP architecture includes a number of IO "Adapters" or event input/output streams. These adapters allow models and agents specified in RML to be integrated into software-based instructional systems. The architecture includes event sources based on soaDM, soaCDO, and Esper. RML behavior models interacting with Esper underlie agent logic. CDOs processed in soaCDO underlie domain knowledge. Finally, semantic networks processed in soaDM underlie declarative memory. Figure 1 illustrates with arrows how these event sources produce events through: (a) the execution of agent logic; (b) the querying of databases containing long-term knowledge; (c) the mining of domain knowledge; and (d) interactions with a large-scale declarative memory. It is these event sources based on agent logic, declarative memory, domain knowledge, and databases that "cognitively enhance" CEP in CECEP. The event cloud (a form of knowledge blackboard or working memory) and pattern matcher (a form of rule engine) in the architecture are technologically realized through Esper.

Models and agents specified in the RML authoring environment are not directly executed. RML specifications are instead translated into executable code artifacts in the following ways:

Declarative Knowledge: is specified as events and relations and processed by code generators that produce files that configure the soaDM semantic network.
Procedural Knowledge: is specified as behavior models and processed by code generators that produce NERML, a text-only DSL formally equivalent to RML that is then translated into Java. Code generation is divided into these steps to support modelers that prefer to specify models and agents directly in NERML.

Fig. 1 The CECEP architecture in which RML models and agents are executed

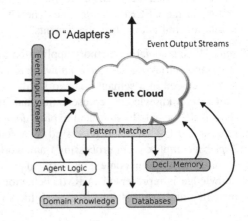

Java files generated from either RML or NERML interact with Esper and govern the behavior of models and agents.

Domain Knowledge: is specified in CDOs and processed by code generators that produce constraint-networks for use in soaCDO.

soaCDO was originally developed in screamer+ [11], a LISP based non-deterministic programming environment that utilizes chronological back tracking algorithms. Screamer+ is fairly efficient in its processing, but carries out evaluations in a serial manner. Complex CDOs can enable powerful autonomous agents, but have large state spaces. These can be slow to search using the serial search algorithms utilized by the screamer+ environment.

This chapter examines how CDOs can be accelerated to enable more powerful autonomous agents by converting CDOs into constraint networks and solve them using a parallel generate and test exhaustive depth first search algorithm on multicore hardware. Two computing platforms were examined: Intel Xeon processors and NVIDIA GPGPUs (general purpose graphical processing units). GPGPUs can have several hundred processing core and thus provide significant speedups over Intel Xeon processors (which typically have less than 10 cores).

3 Constraint Satisfaction Problems in the CECEP Architecture

3.1 Definition of CSPs

A CSP [1–4] consists of 3 components: (1) a set of variables, (2) a finite set of values for each of the variables, and (3) a set of constraints restricting the values that the variables can take simultaneously. A feasible solution to a CSP is found when each variable can be assigned a value that meets the constraints. A CSP could be solved with an objective to find one solution, all solutions, or an approximate solution. Depending on the solution space, a CSP is classified as having a finite or infinite domain.

Finite domain CSPs are solved using systematic search algorithms [12–16]. These algorithms can solve a problem or prove that no solutions exist. Approximate solutions are typically sought for infinite domain CSPs through non-systematic search approaches [3, 13]. These infinite domain searches are typically incomplete and thus cannot guarantee that a solution will be found, or that one exists. The CECEP architecture requires finite domain CSPs because all constraints within the system need to be satisfied, and hence, exact solutions are needed.

3.2 CSP and the Processing of Domain Knowledge Representations in CECEP

RML intelligent agents can effectively use behavior models to trace actions, make sense of data/observations, and determine courses of actions that match goals to situation affordances. Behavior models are particularly effective in task contexts where it is relatively easy to capture anticipated (correct or incorrect) sequences of action. In contexts where multiple actions are appropriate, it can be difficult to specify RML behavior models covering large spaces of alternative actions. Agents executing in the CECEP architecture use CDOs to track and comprehend actions under these circumstances. This section demonstrates how a CDO can be used to capture structural and relational domain knowledge in such a way that constraint-satisfaction processes in soaCDO allow an RML instructional agent to: (1) make sense of actions and intentions; and (2) determine the appropriateness of actions and intentions. Capturing and processing domain knowledge this way in the CE-CEP architecture greatly reduces the burden of behavior model specification.

Figure 2 shows a CDO capturing structural domain knowledge related to a track/aircraft classification task. CDOs capture *structural domain knowledge* in tree-like structures consisting of:

Entities: indicates by gray circles. Entities represent domain constructs. In Fig. 2, entities describe the structural attributes of the set of *track* (or aircraft) entities that are to be classified. Note that in CDOs, the root entity determines the entity set of interest.

Structural Decompositions: indicated by rectangles labeled "and." Decompositions represent fixed entity sub-structures. In Fig. 2, a *track_decomposition* indicates that *track* entities are comprised of *position, movement, ews, model,* and *assessment* sub-entities. Additionally, *assessment* entities are comprised of *threat* and *type* sub-entities.

Choices: indicated by rectangles labeled "xor." Choices represent alternative entity sub-structures. In Fig. 2, *ews_choices, model_choices, threat_choices,* and *type_choices* capture alternative sub-entity choices the trainee will ultimately have to choose between. For example, to classify a track, an agent will likely have to determine its *ews* choice and certainly have to determine its *threat* and type *choices*.

Entity Properties: indicated by attached " ~ " values. Attached values represent entity properties.

CDOs additionally capture *relational domain knowledge* in a constraint language. Domain-specific constraints specified in the constraint language express complex relationships between entities represented in the tree-like structure. Table 1 lists example constraints that capture a sub-set of the classification "rules" governing track/aircraft classification.

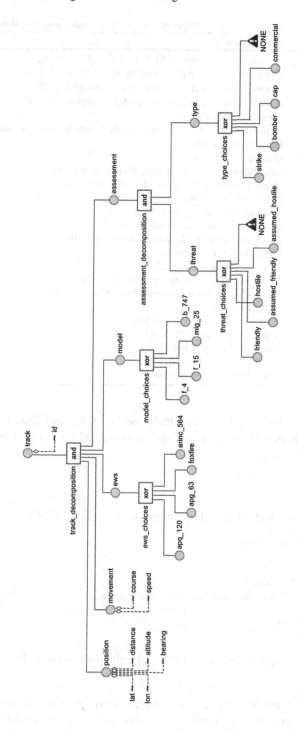

Fig. 2 A cognitive domain ontology capturing knowledge about *track* entities specified in RML

Table 1 Domain-specific constraints integrated into the *track* entities cognitive domain ontology

Name	Specification
C1	iff ews_choices is arinc_564 then model_choices is b_747
C2	iff ews_choices is apq_120 then model_choices is f_4
C3	iff ews_choices is apg_63 then model_choices is f_15
C4	iff ews_choices is foxfire then model_choices is mig_25
C5	if model_choices is f_4
	then threat_choices is assumed_hostile hostile assumed_friendly or friendly
	type_choices is strike
C6	if model_choices is f_15
	then threat_choices is assumed_friendly or friendly
	threat_choices is not (assumed_hostile or hostile)
	type_choices is strike
C7	if model_choices is b_747
	then threat_choices is friendly
	threat_choices is not (assumed_hostile or hostile)
	type_choices is commercial
C8	if model_choices is mig_25
	then threat_choices is assumed_hostile or hostile
	threat_choices is not (assumed_friendly or friendly)
	type_choices is strike
C9	if speed is between 350 and 550,
	altitude is between 25000 and 36000
	then model_choices is b_747
	threat_choices is friendly
	type_choices is commercial

CDOs are transformed into constraint networks by soaCDO. The mapping between CDOs translated into constraint networks and the components of a CSP previously defined are:

(1) CSP variables

    a. "choices" and "attached variables" in a translated CDO,

(2) CSP variable values

    a. "entities" below choices,
    b. "values" that attached variables can take on,

(3) CSP constraints

    a. "constraints" expressing additional relational domain knowledge in a CDO.

These constraint networks are then searched over by a non-deterministic CSP constraint solver. CDO "solutions" are conveyed as sets of events into the CECEP event cloud through an adapter. Choices can be limited through "assertions" that fix the value of a choice or attached variable value. Constraint propagation utilizing assertions can produce solutions that can be exploited by an RML model or agent. For example, when an RML agent is executing in the CECEP architecture, it can use task environment events to assert the values of choices in a CDO. Table 2 shows the impact assertions can have.

Table 2 shows how constraint-based search in soaCDO can determine from observed *speed* and *altitude track* values that a classification agent should assess the *track* as *friendly* and *commercial*. Notice how the bi-directional implication (iff) underlying C1 in Table 1 allows a classification agent to infer (through the exploitation of constraint knowledge) that the *ews_choice* will be *arinc_564*.

In contexts where ambiguity or equally effective actions are possible, it would be virtually impossible to develop an effective RML agent using only behavior models. In such contexts, it is often the case that multiple actions should be considered adequate. Specifying all possible event patterns and event pattern rules under these circumstances would be costly and error prone. Domain knowledge in CDOs significantly decreases the burden of specifying classification agents under these circumstances. Table 3 shows how constraint-based search in soaCDO can determine from observed *ambiguous speed* and *altitude* and *ews_choice track* values that a classification agent can appropriately classify a *track* 4 ways.

Under these circumstances, constraint-based search in soaCDO allows the agent to "mine" courses of action in structural/relational *domain knowledge* in CDOs rather than forward-chaining through *procedural knowledge* in behavior models. In other words, constraint-based search in soaCDO allows the agent designer to develop: (1) a representation of domain knowledge that will allow an agent to determine "what" actions are contextually appropriate; and (2) a simplified set of behavior models that represent "how" to procedurally realize chosen actions. Separation of the "what" and "how" concerns makes it significantly easier to develop agents that manage ambiguity or must consider equally effective actions.

Figure 3 shows a CDO that allows a CECEP agent to exploit inference from evidence to likely causes. Domain-specific constraints similar to those in Table 4 allow the agent to infer from evidence obtained from binary code inspection to expectations about binary behavior. Follow-on "troubleshooting" actions

Table 2 Illustration of how *speed* and *altitude* assertions allow an agent to determine the single correct track/aircraft classification

Assertions	Speed is 500
	altitude is 30000
Solutions	Track
	position { ~ altitude = 30000}
	movement { ~ speed = 500}
	ews {ews_choices = arinc_564}
	model {model_choices = b_747}
	assessment {threat_choices = friendly}, {type_choices = commercial}

Note solution is an abstraction of events returned by soaCDO

Table 3 Illustration of how *speed*, *altitude*, and *ews_choices* assertions allow a tutor agent to determine the courses of action a trainee can undertake when the *threat_choices* aspect of a *track* is ambiguous

Assertions	Speed is 500
	altitude is 15000
	ews_choices is apq_120
Solutions	Track
	position $\{ \sim$ altitude $= 15000\}$
	movement $\{ \sim$ speed $= 500\}$
	ews {ews_choices $=$ apq_120}
	model {model_choices $=$ f_4}
	assessment {threat_choices $=$ hostile}, {type_choices $=$ strike}
	Track
	position $\{ \sim$ altitude $= 15000\}$
	movement $\{ \sim$ speed $= 500\}$
	ews {ews_choices $=$ apq_120}
	model {model_choices $=$ f_4}
	assessment {threat_choices $=$ friendly}, {type_choices $=$ strike}
	Track
	position $\{ \sim$ altitude $= 15000\}$
	movement $\{ \sim$ speed $= 500\}$
	ews {ews_choices $=$ apq_120}
	model {model_choices $=$ f_4}
	assessment {threat_choices $=$ assumed_friendly}, {type_choices $=$ strike}
	Track
	position $\{ \sim$ altitude $= 15000\}$
	movement $\{ \sim$ speed $= 500\}$
	ews {ews_choices $=$ apq_120}
	model {model_choices $=$ f_4}
	assessment {threat_choices $=$ assumed_hostile}, {type_choices $=$ strike}

producing additional evidence from a targeted analysis of binary behavior allow the agent to strengthen or weaken hypotheses about likely *binary_classification*.

Agents similar to the one just discussed can exploit declarative, procedural, and domain knowledge in the CECEP architecture while monitoring, classifying, and making sense of complex patterns of events in an event cloud. Adapters can integrate packet sniffers, instrumented reverse engineering tools, and interaction with humans into the CECEP architecture. CECEP agents could have a substantial impact as monitoring agents in net-centric computing infrastructures or reverse engineering tools. To have these impacts, CECEP agents must be able to represent and mine large and complex CDOs. To ensure the feasibility of mining of large CDOs, a high-performance CSP capability must be developed and incorporated into soaCDO. The next section describes an effort to do precisely this.

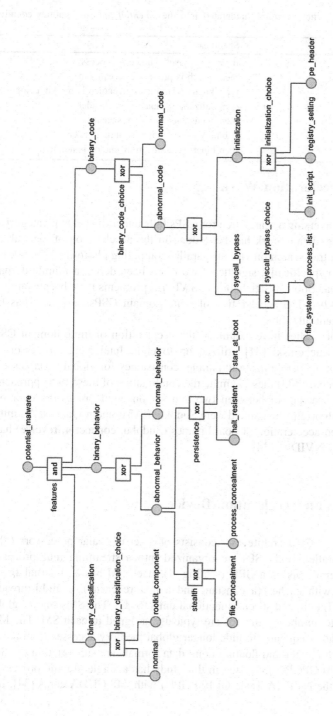

Fig. 3 A cognitive domain ontology capturing knowledge about *potential_malware*

Table 4 Domain-specific constraints integrated into the *potential_malware* entities cognitive domain ontology

Name	Specification
C1	if syscall_bypass_choice is process_list
	then stealth is process_concealment
	type_binary_classification_choice is rootkit_component
C2	if initialization_choice is pe_header
	then persistence is halt_resistent
C3	if binary_code_choice is normal_code
	then binary_classification_choice is nominal

4 Related Acceleration Work

The process of traversing a search tree for CSPs falls under the general category of graph search. Significant work has been done in the acceleration of generalized graph search and tree search on various parallel computing platforms [17–24]. For CSP acceleration specifically, significant work has been done on infinite domain CSPs; in particular Boolean satisfiability (SAT) [2] problems have been examined extensively [25–29]. The acceleration of finite domain CSPs, however, has not received significant attention.

Several recent studies have examined the acceleration of finite domain CSPs [30]. Rolf and Kuchcinski [31] utilized up to eight Intel Xeon processors to explore two forms of parallelization: parallel consistency for global variables [32] and parallel search. In [33] they examine the combination of these two approaches and achieved a speedup of between two and six on eight processors over one processor. The algorithms examined were Sudoku, LA31, and n-Queens. A limited set of GPU based acceleration of board games (Sudoku, connect-4, reverse) have been released by NVIDIA [34].

5 GPGPU as an Acceleration Device

NVIDIA's CUDA GPU architecture consists of a set of scalar processors (SPs) operating in parallel. Eight SPs are organized into a streaming multiprocessor (SM), with several SMs on a GPU (see Fig. 4). Each SM has an internal shared memory along with caches (a constant and a texture cache), a multi-threaded instruction (MTI) unit, and special functional units (SFU). The SMs share a global memory. A large number of threads are typically assigned to each SM. The MTI can switch threads frequently to hide longer global memory accesses from any of the threads. Both integer and floating point data formats are supported in the SPs. The Tesla C2070 GPGPU [35] used in this study has a single graphic processing chip (based on the NVIDIA Tesla GF100 GPU) with 448 CUDA cores [34]. It is capable of 1.03 TFLOPs of processing performance and comes standard with 6 GB of GDDR5 memory at 144 GB/s bandwidth.

In the CUDA programming language, tasks are divided between the CPU host and GPGPU device. Typically the host code calls the GPGPU device code (called the kernel). To be effective a kernel should have thousands of threads distributed across the CUDA cores on the GPGPU. Threads are lightweight, with low switching costs. Each CUDA core runs a single thread at a time and switches to other threads whenever there is a memory access delay. Running thousands of threads on the GPGPU allows cores to hide memory access latencies by switching to other threads that are not waiting on memory accesses. In the NVIDA CUDA environment, identical threads are grouped into blocks, while a group of identical blocks forms a grid. In the systems studied, a CUDA kernel supports 1 grid only. Thus a kernel can have a very large number of identical threads running under it, leading to massive amounts of parallelism.

6 Hardware Acceleration of CDOs

In order to accelerate CDOs through specialized parallel hardware, it is necessary to convert a CDO into an equivalent constraint network for solving. The searching of this constraint network based on the input constraints provides the solution sought. This section starts by examining how a CDO can be converted to a CSP

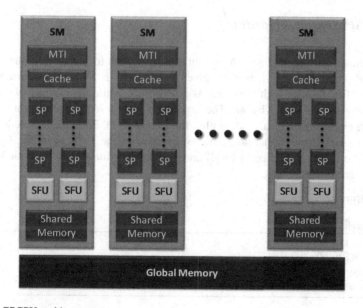

Fig. 4 GPGPU architecture

and then examines how the search of the CSP can be parallelized on to specialized parallel hardware.

6.1 Representing CDOs as CSPs

In this study, a CDO is represented as an equivalent tree structure that can be searched through a CSP. Consider the example CDO in Fig. 5 describing a set of ball entities. Figure 6 shows a multi-layered tree representation of this CDO, with each layer corresponding to a specialization. The total number of leaf nodes in this tree represents all possible types of ball entities. For instance, the node labeled 14 is a brown colored, medium sized football.

A user would typically search for one or more solutions from a CDO based on a set of constraints. For the CDO in Fig. 5, a set of constraints could be:

If *sport* is *baseball*, then *size* is *small* and *color* is *white*.
If *sport* is *football*, then *size* is *medium* and *color* is *brown*.
If *sport* is *basketball*, then *size* is *large* and *color* is *orange*.

Finding the solution to the CDO can be treated as solving a constraint satisfaction problem to determine which leaf node(s) within the search tree in Fig. 6 satisfy the user constraints. An exhaustive depth first search algorithm was used to check each leaf node for a match to the user constraints.

6.2 CSP Search Algorithm

An exhaustive depth first search algorithm was written to search for solutions to the CSP within the search tree, as given by Algorithm 1. In this algorithm the number of leaf nodes within the tree is given by N, while the number of choices within the tree is given by M. The algorithm makes N iterations, where each iteration evaluates one leaf node within the tree (line 1). For each leaf node i, the choices associated with that node are first generated (line 2). Thus in Fig. 6, when i is 14, the values of choices(1 to M) are {football, medium, brown}, with $M = 3$ as there are three choices.

Algorithm 1

CSP search algorithm

---
1. For $i = 1$ to N
2. *choices*(1 to M) = *generate_choice_values*(i);
3. If (constraints_satisfied(*choices*(1 to M), *constraints*(1 to P)) == TRUE)
4. Add i to *solution_space*;

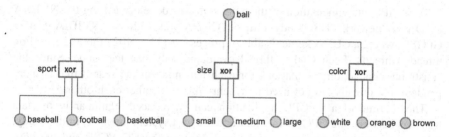

Fig. 5 CDO representing properties of sports balls

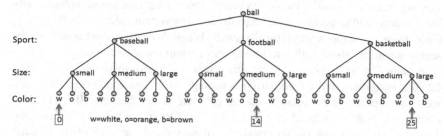

Fig. 6 Tree representing different combinations of properties in the sports balls CDO in Fig. 5

Once choice values are generated, they are compared against each of the constraints provided (line 3). In Algorithm 1, the number of constraints is given by P. If the choices associated with the leaf node under evaluation satisfy all the constraints, then the index for that node is added to the *solution_space* list (line 4).

A C implementation of the algorithm was developed and tested on an Intel Xeon processor. The output of the C program was verified to be correct by comparing to a screamer+ evaluation of the CDO (as in Fig. 5). The C program was about 30 % faster than the screamer+ runtime.

6.3 Parallelizing Search

Each iteration in Algorithm 1 evaluates whether a leaf node within a search tree (such as Fig. 6) satisfies all the constraints provided. This evaluation can be carried out independently for each leaf node. A multi-threaded C implementation of the algorithm was developed on a six core Intel Xeon X5650 2.66 GHz processor, where the leaf nodes in the tree were distributed across the threads. The POSIX thread library was utilized in this code. As shown in the results section, large trees gave a linear speedup based on the number of active threads within the system.

A GPGPU implementation of the algorithm was developed using the NVIDIA CUDA framework. This study utilized CUDA 4.0 with the NVIDIA C2070 GPGPU, where a CUDA kernel is able to process 65535×1024 (about 67 million threads) threads. Each CUDA thread evaluated only one leaf node. Since the constraint satisfaction calculations for each leaf node in the search tree is independent, the threads did not need to synchronize with other each other.

The runtime of a GPGPU application can be reduced significantly by data transfers between the GPGPU memory and the host system memory. In this study, the amount of data transfers was minimal because the CDO/CSP and the constraints could be described by small data structures.

These data items needed to be transferred only once into the GPGPU global memory before the start of the search kernel. Once execution was completed, only the solutions had to be transferred back to the host system memory. In this study, the number of solutions was typically small. When there were more solutions, the search would be aborted after a limited set of solutions were found.

A multi-GPGPU version of the program was developed where the host for each GPGPU communicated with other hosts through MPI. The search tree leaf nodes were distributed evenly amongst all the compute nodes through MPI. This was carried out on the Condor cluster at the Air Force Research Lab (ARFL/RI) [36]. This is a heterogeneous cluster consisting of 84 servers, each with two NVIDIA GPGPUs and 1716 Sony Playstation 3 based Cell processors. This study utilized the C2070 GPGPUs on this cluster for the MPI based studies.

7 Results

To examine the impact of CDO complexity on runtime, five synthetic CDOs of varying complexities were developed as shown in Table 5. Each CDO was similar to the CDO in Fig. 5, except that the number of choices and entities under each choice was changed. Five CDOs with 8 to 12 choices under the root entity were developed. The number of entities under each choice was made the same as the number of choices in the CDO. Thus the CDO in Fig. 5 has three choices, and each choice has three entities under it. Table 5 shows the properties of the CSP trees

Table 5 Synthetic CDO's and CSP's Examined

CDO		CSP tree	
Choices	Entities under each choice	Tree levels	Leaf Nodes
8	8	8	2×10^7
9	9	9	4×10^8
10	10	10	1×10^{10}
11	11	11	3×10^{11}
12	12	12	9×10^{12}

Table 6 Run time of search trees on different platforms

Tree levels	Constraints	1 Intel xeon core	8 Intel xeon cores	1 GPU	4 GPUs	8 GPUs
10	10	2512 s (42 min)	321 s (5 min 2 s)	20 s	5.7 s	3 s
11	11	79,144 s (22 h)	9720 s (2.7 h)	628 s (10 min 28 s)	158 s	79 s
12	12	2,592,000 s* (30 days)	328,320 s* (3.8 days)	20,721 s (5.75 h)	5400 s (1.5 h)	2700 s (45 min)

* Estimated CPU run time

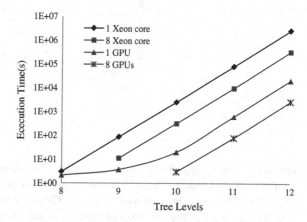

Fig. 7 Run time comparison of CPU, multi-CPU and one GPU

corresponding to these CDOs. These CSP trees have the same number of levels as choices in their corresponding CDOs. As shown in Table 5, the number of leaf nodes increases exponentially with the number of choices in the CDO.

Table 6 and Fig. 7 show the runtime for searching through each entire CSP tree in Table 5. The number of constraints for each case was set to the number of tree levels. The 8 Xeon cores were utilized through multi-threading, while the 8 GPGPUs were used through MPI.

Table 5 shows that the number of leaf nodes increases logarithmically with the tree size. This causes the logarithmic increase in runtimes seen in Fig. 7. For trees with 10 levels, the execution time on one Xeon core was 2,512 s (approximately 42 min) and for trees with 11 levels it became 79,144 s (approximately 22 h). An increase of just one level in the tree increased the computation time by almost 30 times. The Xeon runtime of the tree with 12 levels was estimated based on the runtime of the smaller trees, as the actual runtimes were too long to run on computers available.

Table 7 shows the speedup of the different parallel configurations in Table 6. The multicore version of the code provided a speedup roughly equivalent to the

Table 7 Speedups over single CPU (based on data in Table 6)

Tree levels	8 Intel xeon cores	1 GPU	4 GPUs	8 GPUs
10	7.8	125.6	440.7	837.3
11	8.1	126.0	500.9	1001.8
12*	7.9	125.1	480.0	960.0

* Speedup over estimated single CPU run time

Fig. 8 Comparison of experimental results versus estimated run times from Eq. (1)

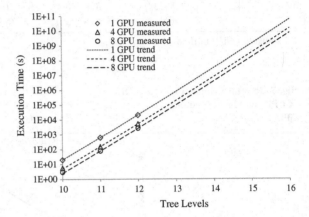

number of cores used (8 cores), indicating that the search process utilized is high parallelizable. A single GPGPU provided speedups of over 125 times the Xeon processor. The speedup with four and eight GPGPUs increased slightly with tree size. This is likely to the overhead of MPI communications having a greater impact on the smaller tree.

Given that the runtimes in Table 6 increase linearly with the number of nodes within the search tree, a simple model can be developed to predict the runtimes for larger trees. Eq. (1) represents the runtime in terms of the time to compute one constraint on a GPU, the number of constraints provided by the user, the number of nodes in the CSP search tree, and the number of GPUs utilized.

$$\text{Execution time} = \frac{(\text{Number of nodes in tree}) \times (\text{Number of constraint}) \times (\text{Time required to compute one constraint})}{\text{Number of GPUs}}$$

(1)

The results of this equation are plotted in Fig. 8 for different tree levels using 1, 4, and 8 GPUs. The results show that the measured data fall along the trend lines from Eq. (1). The plot shows that the run time would increase exponentially with problem size as expected. This makes a strong case of reducing the search space through tree pruning appoarches.

8 Conclusion

This chapter introduced a cognitively enhanced complex event processing (CE-CEP) architecture being developed in the Air Force Research Laboratory (AFRL). CECEP enables computer aided decision making through the mining of domain knowledge. The cognitive domain ontology (CDO) data structure within this architecture can require extremely long computational run times to search using the situational constraints seen by an agent. This study examined the parallelization of this search process onto multicore processors and GPGPU clusters. Speedups of almost a 1000 times were seen using eight NVIDIA Tesla C2070 GPGPUs over one Xeon X5650 processor core. Speedups of this level will allow more complex CDOs to be examined in real time on reactive agents and thus enable systems with enhanced intelligence to be designed.

References

1. A. Aiken, Introduction to set constraint-based program analysis. Sci. Comput. Program. **35**, 79–111 (1999)
2. A.K. Mackworth, Constraint Satisfaction, in *Encyclopedia of Artificial Intelligence*, ed. by S.C. Shapiro, (John Wiley and Sons, New York, 1987)
3. S.C. Brailsford, C.N. Potts, B.M. Smith, Constraint satisfaction problems: algorithms and applications. Eur. J. Oper. Res. **119**, 557–581 (1999)
4. S. Russell, P. Norvig, *Artificial Intelligence: A Modern Approach*, 3rd edn. (Prentice Hall, 2009)
5. S. Douglass, C. Myers, Concurrent knowledge activation calculation in large declarative memories, in *Proceedings of the 10th International Conference on Cognitive Modeling*, 2010, pp. 55–60
6. S. Douglass, S. Mittal, Using domain specific languages to improve scale and integration of cognitive models, in *Proceedings of the Behavior Representation in Modeling and Simulation Conference*, Utah, USA, 2011
7. Z. Molnár, D. Balasubramanian, A. Lédeczi, An introduction to the GenericModeling Environment, in *Proceedings of the TOOLS Europe 2007 Workshop on Model-Driven Development Tool Implementers Forum* (Zurich, Switzerland, 2007)
8. A. Ledeczi, M. Maroti, A. Bakay, G. Karsai, J. Garrett, C. Thomason, G. Nordstrom, J. Sprinkle, P. Volgyesi, *"The generic modeling environment,"* Workshop on Intelligent Signal Processing, vol. 17, (Budapest, Hungary, 2001)
9. A. Ledeczi, P. Volgyesi, G. Karsai, Metamodel composition in the Generic Modeling Environment. Comm. at Workshop on Adaptive Object-Models and Metamodeling Techniques, Ecoop, vol. 1, 2001
10. Esper available: http://esper.codehaus.org/
11. J.M. Siskind, *Screaming Yellow Zonkers* (MIT Articial Intelligence Laboratory, 1991)
12. P. Sanders, Better Algorithm for Parallel Backtracking, in *Workshop on Algorithms for Irregularly Structured Problems*, number 980 in LNCS, 1995
13. E.C. Freuder, R. Dechter, M.L. Ginsberg, B. Selman, E.P.K. Tsang, Systematic versus stochastic constraint satisfaction, in *Proceedings of the 14th International Joint Conference on Artificial Intelligence*, vol. 2 (Morgan-Kaufmann, 1995), pp. 2027–2032

14. M. Ginsberg, D. McAllester, GSAT and Dynamic Backtracking, in Proceedings of the Fourth Int'l Conf. Principles of Knowledge Representation and Reasoning, 1994, pp. 226–237

15. I. Lynce, L. Baptista, J.P. Marques-Silva, Stochastic Systematic Search Algorithms for Satisfiability, in *The LICS Workshop on Theory and Apps of Satisfiability Testing*, 2001

16. J.R. Bitner, E. Reingold, Backtracking programming techniques. Commun. ACM **18**(11), 651–656 (1975)

17. L.L. Wong, M.W.-M. Hwu, An effective GPU implementation of breadth-first search, Design Automation Conference (DAC), 2010, pp. 52–55

18. D. Sulewski, *Large-Scale Parallel State Space Search Utilizing Graphics Processing Units and Solid State Disks*, Dissertation, Dortmund University of Technology, 2011

19. J. Jenkins, I. Arkatkar, J.D. Owens, A. Choudhary, N.F. Samatova, Lessons learned from exploring the backtracking paradigm on the GPU, in *Euro-Par 2011: Proceedings of the 17th International European Conference on Parallel and Distributed Computing*, Lecture Notes in Computer Science, vol. 6853 (Springer, August/September 2011), pp. 425–437

20. A. Buluc, K. Madduri, Parallel Breadth First Search on Distributed Memory Systems, in *The International Conference for High Performance Computing*, Networking, Storage and Analysis, 2011

21. D. Merrill, M. Garland, A. Grimshaw, Scalable GPU Graph Traversal, in *Proceedings of PPoPP*, February 2012

22. P. Harish, P. Narayanan, Accelerating large graph algorithms on the GPU using CUDA, in *High Performance Computing—HiPC 2007: 14th International Conference, Proceedings*, ed. by S. Aluru, M. Parashar, R. Badrinath, V. Prasanna. vol. 4873 (Springer-Verlag, Goa, India, 2007), pp. 197–208

23. S. Hong, S. Kim, T. Oguntebi, K. Olukotun, Accelerating CUDA Graph Algorithms at Maximum Warp, in *Proceedings of the 16th ACM symposium on Principles and practice of parallel programming*, 2011

24. S.D. Joshi, V.S. Inamdar, Performance improvement in large graph algorithms on GPU using CUDA: an overview. Int. J. Comp. Appl. **10**(10), 10–14 (2010)

25. Y. Wang, NVIDIA CUDA Architecture-based Parallel Incomplete SAT Solver, Master Project Final Report, Rochester Institute of Technology, 2010

26. P. Leong, C. Sham, W. Wong, W. Yuen, M. Leong, A bitstream reconfigurable FPGA implementation of the WSAT algorithm. IEEE Trans. VLSI Syst. **9**(1), 197–201 (2001)

27. D. Diaz, S. Abreu, P. Codognet, Parallel constraint-based local search on the Cell/BE multicore architecture, in *Intelligent Distributed Computing IV. Studies in Computational Intelligence*, vol. 315, ed. by M. Essaaidi, M. Malgeri, C. Badica (Springer, Heidelberg, 2010), pp. 265–274

28. D. Diaz, S. Abreu, P. Codognet, Targeting the Cell Broadband Engine for Constraint-Based Local Search. (Published Online October 20, 2011). doi: 10.1002/cpe.1855

29. Y. Caniou, P. Codognet, D. Diaz, S. Abreu, Experiments in parallel constraint-based local search, in *EvoCOP'11, 11th European Conference on Evolutionary Computation in Combinatorial Optimisation*. Lecture Notes in Computer Science (Springer Verlag, Torino, Italy, 2011)

30. I.P. Gent, C. Jefferson, I. Miguel, N.C.A. Moore, P. Nightingale, P. Prosser, C. Unsworth, A Preliminary Review of Literature on Parallel Constraint Solving, Computing Science, Scotland Workshop on Parallel Methods for Constraint Solving (Glasgow and St. Andrews Universities, 2011)

31. C. Rolf, *Parallelism in Constraint Programming*, Ph.D. thesis, 2011

32. C. Rolf and K. Kuchinski, Parallel Consistency in Constraint Programming. The International Conference on Parallel and Distributed Processing Techniques and Applications: SDMAS Workshop, 2009

33. C. Rolf and K. Kuchinski, Parallel Search and Parallel Consistency in Constraint Programming. International Conference on Principles and Practices of Constraint Programming, 2010

34. GPU AI for Board Games, http://developer.nvidia.com/gpu-ai-board-games, Accessed 10 July 2012
35. NVIDIA Tesla C2070. http://www.nvidia.com/docs/IO/43395/BD-04983-001_v05.pdf
36. M. Barnell, Q. Wu, R. Luley, Integration and development of the 500 TFLOPS heterogeneous cluster (Condor). IEEE High Perform. Extreme Comput. Conf. 2012

Memristors and the Future of Cyber Security Hardware

Michael J. Shevenell, Justin L. Shumaker, Arthur H. Edwards
and Robinson E. Pino

1 Introduction

Much of today's cyber defense infrastructure exists as software executing on various
forms of digital hardware. Historically speaking this approach has been adequate,
yet it is widely acknowledged that the gap between new data and available processing power is cause for great concern. Many different hardware acceleration
technologies have been successfully employed over the years to address this problem, these include programmable logic devices, graphics processors, vectorized
instruction sets as well as multi-core and distributed processing architectures. One
technology that is poised to narrow this gap is the memristor, a two-terminal analog
memory device. After the devices first tangible appearance in 2008 researchers have
identified several key areas in which memristors will have a significant impact with
cyber security being one. In the near term memristors will be utilized as binary
storage devices, whose performance will rival flash memory technology. It is also
very likely that memristive programmable logic gates will outperform existing
CMOS technologies. However, the most significant contribution from the memristor
is likely to be the exploitation of what is known as the nonlinear device region.

When used as binary devices, memristors are programmed using "set" and
"reset" pulses. These "set" and "reset" pulses form low and high resistance states

M. J. Shevenell (✉) · R. E. Pino
ICF International, Baltimore, MD, USA
e-mail: Michael.Shevenell@icfi.com

R. E. Pino
e-mail: Robinson.Pino@icfi.com

J. L. Shumaker
Army Research Laboratory, Aberdeen, MD, USA
e-mail: justin.l.shumaker.civ@mail.mil

A. H. Edwards
Air Force Research Laboratory, Albuquerque, NM, USA
e-mail: Arthur.Edwards@kirtland.af.mil

R. E. Pino (ed.), *Network Science and Cybersecurity*,
Advances in Information Security 55, DOI: 10.1007/978-1-4614-7597-2_17,
© Springer Science+Business Media New York 2014

respectively. The device may be "on", "off" or somewhere in-between. Regardless of the whether the device is bipolar, unipolar or nonpolar each has a voltage threshold such that when it is exceeded a change in resistance occurs. This is known as the Nonlinear Device Region. Once the resistance is altered it remains in that state, hence it has "memristance". Unfortunately, the NDR is rather chaotic and cannot be captured by a simple low order curve fit. As research matures on characterizing the NDR for different memristor materials it will become possible to exploit memristance to its full potential. The memristance resolution is therefore a function of material properties and the model governing its control. Many different material configurations of been derived in the development of these Metal–Insulator/Oxide-Metal devices, each exhibiting unique properties. Switching speed, resistance ratios and NDR behavior vary among the many material implementations. The memristor therefore offers a multitude of capabilities for a new class of analog computing device hardware where the stored value does not require energy to maintain its integrity.

Researchers are now designing new analog computing architectures that exploit the NDR in a manner that compliments existing CMOS technology for the purpose of increasing performance and reducing power consumption. This is analogous to the manner in which PGA's are coupled with central processing units to speed up parallel algorithms. A memristive computing architecture will have major implications toward improving the performance of cyber threat detection. One of the envisioned computer architectures for performing threat detection draws from our knowledge of the brain and its complex architecture. The brain is very good at performing associations and making predictions. The cerebral cortex and neocortex are largely responsible for these functions. As the functional understanding of these regions in the brain improves it will lead to higher fidelity neural network architectures. This does not mean the system will be intelligent, but should provide for a more efficient computing architecture.

One concept that has been widely publicized about memristors is their ability to mimic the behavior of a biological synapse. The strengthening and weakening of the synapse is the result of an electrochemical response resulting from the time differential in pre and post synaptic neuronal firing known as Spike Time Dependent Plasticity. It has been shown, with varying success, that STDP neural networks are able to function as a dynamically reconfigurable hardware able to adapt to changing stimuli in situ. It is this type of dynamic hardware that is necessary to meet the challenges of modern cyber threats. Regardless of the training algorithm or network architecture it is important to note that a neural network is an analog process in nature and should exist on analog hardware. While analog Application Specific Integrated Circuits that mimic specific mechanisms within the brain exist, none are utilizing the memristor as the fundamental computing element commercially.

In order to transition from digital processor based computing to analog memristor based computing researchers have many fundamental challenges to overcome. Currently, a memristor based computer is not in existence; however researchers are developing systems to emulate the neuromorphic parallel computing architectures envisioned for memristor technology. The first challenge is to develop a suitable research framework to begin emulation experimentation and characterization of a memristor based computer. The emulation of the memristor computer must employ the following fundamental characteristics.

Memristor Computer Fundamental Characteristics:

- Learning: A memristor computer must be trained to learn an internal representation of the problem. No algorithm is needed.
- Generalization: Training the memristor computer with suitable samples.
- Associative Storage: Information is stored on the memristor according to its content.
- Distributed Storage: The redundant information storage is distributed over all memristor neurons.
- Robustness: Sturdy behavior in the case of disturbances or incomplete inputs.
- Performance: Efficient massive parallel structure.

Researchers have engineered a range of frameworks which utilizes approaches and techniques to satisfy the fundamental characteristics of a memrister computer. The various frameworks have made successful attempts to emulate the memristor computer by utilizing the neural network approach.

2 Basic Memristor Computer Neural Network Approach

Computers today can perform complicated calculations, handle complex control tasks and store huge amounts of data. However, there are classes of problems which a human brain can solve easily, but a computer can only process with a high computational power. Examples are character recognition, cognitive decision making, image interpretation and network intrusion detection. The class of problems suitable for emulating human activity is also suitable for the memristor computer. One area of research investigation of the memrister computer is to apply neural networks to network and host-based intrusion detection.

Network based intrusion detection attempts to identify unauthorized, illicit, and anomalous behavior based solely on network traffic. A network IDS, using either a network tap, span port, or hub collects packets that traverse a given network. Using the captured data, the IDS system processes and flags any suspicious traffic. Unlike an intrusion prevention system, an intrusion detection system does not actively block network traffic. The role of a network IDS is passive, only gathering, identifying, logging and alerting.

Host based intrusion detection attempts to identify unauthorized, illicit, and anomalous behavior on a specific device. HIDS generally involves an agent installed on each system, monitoring and alerting on local OS and application activity. The installed agent uses a combination of signatures, rules, and heuristics to identify unauthorized activity. The role of a host IDS is passive, only gathering, identifying, logging, and alerting.

The most important property of a property of a Neural Network is to automatically learn coefficients in the Neural Network according to data inputs and outputs. When applying the Neural Network approach to Intrusion Detection, we first have to expose the neural network to normal data and to network attacks. Next, we automatically adjust the coefficients of the neural network during the training phase. Performance tests are then conducted with real network traffic and attacks to determine the detection rate of the learning process.

Neural network can be used to learn users behavior. A neural network can learn the predictable behavior of a user and fingerprint their activities. If a user's behavior does not match his activities, the system administrator can be alerted of a possible security breech.

Unlike computers, the human brain can adapt to new situations and enhance its knowledge by learning. It is capable to deal with incorrect or incomplete information and still reach the desired result. This is possible through adaption. There is no predefined algorithm, instead new abilities are learned. No theoretical background about the problem is needed, only representative examples.

The neural approach is beneficial for the above addressed classes of problems. The technical realization is called neural network or artificial neural network (ANN). They are simplified models of the central nervous system and consist of intense interconnected neural processing elements. The output is modified by learning. It is not the goal of neural networks to recreate the brain, because this is not possible with today's technology. Instead, single components and function principles are isolated and reproduced in neural networks.

The traditional problem solving approach analyses the task and then derives a suitable algorithm. If successful, the result is immediately available. Neural networks can solve problems which are difficult to describe in an analytical manner. But prior to usage, the network must be trained.

Biological neural systems are an organized and structured assembly of billions of biological neurons. A simple biological neuron consists of a cell body which has a number of branched protrusions, called dendrites, and a single branch called the axon as shown in Fig. 1. Neurons receive signals through the dendrites from neighboring connected neurons [1]. When these combined excitations exceed a certain threshold, the neuron fires an impulse which travels via an axon to the other connected neurons. Branches at the end of the axons form the synapses which are connected to the dendrites of other neurons. The synapses act as the contact between neurons and can be excitatory or inhibitory. An excitatory synapse adds to the total of signals reaching a neuron and an inhibitory synapse subtracts from this total. Although this description is very simple, it outlines all those features which are relevant to the modeling of biological neural systems using artificial neural networks. Generally, artificial neuron models ignore detailed emulation of biological neurons and can be considered as a unit which receives signals from other units and passes a signal to other units when its threshold is exceeded. Many of the

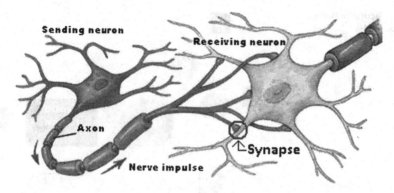

Fig. 1 Synapse of interconnection neurons [1]

key features of artificial neural network concepts have been borrowed from biological neural networks. These features include local processing of information, distributed memory, synaptic weight dynamics and synaptic weight modification by experience. An artificial neural network contains a large number of simple neuron-like processing units, called neurons or nodes along with their connections. Each connection generally "points" from one neuron to another and has an associated set of weights [2, 3].

- Dendrites: Carry electric signals from other cells into the cell body
- Cell Body: Sum and threshold the incoming signals
- Axon: Signal transfer to other cells
- Synapse: Contact point between axon and dendrites
- Neurons: The neuron in neural networks is the equivalent to nerve cells in the central nervous system.

Neural networks are models of biological neural structures. The starting point for most neural networks is a model neuron, as in Fig. 2. This neuron consists of multiple inputs and a single output. Each input is modified by a weight, which multiplies with the input value. The neuron will combine these weighted inputs and, with reference to a threshold value and activation function, use these to determine its output. This behavior follows closely our understanding of how real neurons work.

While there is a fair understanding of how an individual neuron works, there is still a great deal of research and mostly conjecture regarding the way neurons organize themselves and the mechanisms used by arrays of neurons to adapt their behavior to external stimuli. There are a large number of experimental neural network structures currently in use reflecting this state of continuing research.

In our case, we will only describe the structure, mathematics and behavior of that structure known as the back propagation network. This is the most prevalent and generalized neural network currently in use.

Fig. 2 Neural network neuron model

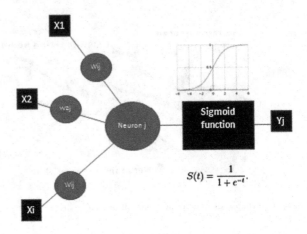

Fig. 3 Neural network layers

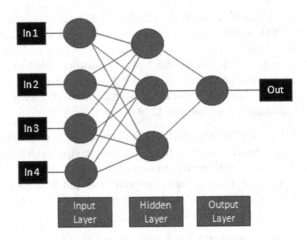

   To build a back propagation network, proceed in the following fashion. First, take a number of neurons and array them to form a layer [4, 5]. A layer has all its inputs connected to either a preceding layer or the inputs from the external world, but not both within the same layer. A layer has all its outputs connected to either a succeeding layer or the outputs to the external world, but not both within the same layer.

   Next, multiple layers are then arrayed one succeeding the other so that there is an input layer, multiple intermediate layers and finally an output layer, as in Fig. 3. Intermediate layers, that is those that have no inputs or outputs to the external world, are called > hidden layers. Back propagation neural networks are usually fully connected. This means that each neuron is connected to every output from the preceding layer or one input from the external world if the neuron is in the first layer and, correspondingly, each neuron has its output connected to every neuron in the succeeding layer [2].

Generally, the input layer is considered a distributor of the signals from the external world. Hidden layers are considered to be categorizers or feature detectors of such signals. The output layer is considered a collector of the features detected and producer of the response. While this view of the neural network may be helpful in conceptualizing the functions of the layers, you should not take this model too literally as the functions described may not be so specific or localized.

Learning in a neural network is called training. One of the basic features of neural networks is their learning ability. To obtain the expected result, the network must reach an internal representation of the problem. Like training in athletics, training in a neural network requires a coach, someone that describes to the neural network what it should have produced as a response. From the difference between the desired response and the actual response, the error is determined and a portion of it is propagated backward through the network. At each neuron in the network the error is used to adjust the weights and threshold values of the neuron, so that the next time, the error in the network response will be less for the same inputs.

3 Learning Methods are Subdivided into two Classes

3.1 Supervised Learning

The network is trained with samples of input–output pairs. The learning is based on the difference between current and desired network output.

3.2 Unsupervised Learning

The network is only trained with input samples, the desired output is not known in advance. Learning is based on self-organization. The network autonomously divides the input samples into classes of similar values

4 Emulation of the Memristor Computer

The memristor computer emulated neural network approach can be divided into three basic deployment methods. The first basic deployment method uses software based algorithms installed onto the traditional von Neumann CPU architecture (VNCA) using x86 CPUs second, software based algorithms deployed on the VNCA utilizing a Graphics Processing Units (GPUs) and third, a hardware architecture deployed onto a field-programmable gate array (FPGA).

5 Traditional Von Neumann CPU Architecture Approach

The most common deployment method is using neural networks on Linux systems using multiple x86 CPUs or a cluster of computers which execute custom neural network application software. The multiple CPU configuration is usually one system and the cluster is usually configured in this manner. The components of a cluster are usually connected to each other through fast local area networks, each node (computer used as a server) running its own instance of an operating system.

One of the important objectives of the CPU or Cluster approach is to parallelize the training of the neural network by using a VNAC approach is to take advantage of multiple systems and CPUs. The approach uses neural network simulators which are software applications that simulate the behavior of artificial or biological neural networks. They focus on one or a limited number of specific types of neural networks. They are typically stand alone and not intended to produce general neural networks that can be integrated in other software. Simulators usually have some form of built-in visualization to monitor the training process. Some simulators also visualize the physical structure of the neural network.

Besides the hardware as basic condition for any parallel implementation, the software has to be considered as well. Parallel programming must take the underlying hardware into account. First, the problem has to be divided into independent parts which can later be processed in parallel. Since this requires a rather deep understanding of the algorithm, automatic routines to parallelize the problem based on an analysis of data structures and program loops usually lead only to weak results. Some compilers of common computer languages offer this option. In most cases a manual parallelization still offers more satisfying results. Fortunately neural networks provide originally a certain level of parallelism as already mentioned above.

Commonly used mathematical or technical computer languages (C, C++, Fortran) are also available on parallel computers, either with specialized compilers or with particular extensions to code instructions controlling the parallel environment. Using a parallelizing compiler makes working not very different from a sequential computer. There are just a number of additional instructions and compiler options. However, compilers that automatically parallelize sequential algorithms are limited in their applicability and often platform or even operating system dependent. Obviously, the key to parallel programming is the exchange or distribution of information between the nodes. The ideal method for communicating a parallel program to a parallel computer should be effective and portable which is often a conflict. A good compromise is the Message Passing Interface (MPI) which was originally designed to be used with homogeneous computer clusters (Beowulf). It complements standard computer languages with information distribution instructions. Since it is based on C or Fortran and its implementation is pretty effective and available on almost all platforms and operating systems, it has evolved into the probably most frequently used parallel programming language [6].

In case of a heterogeneous computer cluster a similar system the Parallel Virtual Machine (PVM) is widespread and has become the de facto standard. It was developed to provide a uniform programming environment for computer clusters consisting of different nodes running possibly different operating systems, which are considered to be one virtual parallel computer. Since real parallel computers and homogeneous clusters are a subgroup of heterogeneous clusters, PVM is also available on these systems. Two additional parallel programming environments which have similar features as PVM are Pthreads and OpenMP.

6 The GPU Approach

The graphics processing unit (GPU) configuration can also be implemented on the VNCA approach on both the single system multiple CPU or the cluster environment. The GPU is a specialized computer graphics card. The GPU is designed to rapidly manipulate and alter memory to accelerate the building of images in a frame buffer intended for output to a display. Modern GPUs are very efficient at manipulating computer graphics, and their highly parallel structure makes them more effective than general-purpose CPUs for algorithms where processing of large blocks of data is done in parallel. This parallel computing capability make them well suited for implementing neural network algorithms [7, 8, 9]. Since a Neural Network requires a considerable number of vector and matrix operations to get results, it is very suitable to be implemented in a parallel programming model and run on a GPU [10].

The reason memristor computer emulation using a neural network is suitable for GPU is that the training and execution of a neural network are two separate processes. Once properly trained, no writing access is required while using a neural network. Therefore, there is no synchronization issue that needs to be addressed. Moreover, neurons on a same network level are completely isolated, such that neuron value computations can achieve highly parallelization.

To successfully take advantage of the GPU, applications and algorithms should present a high degree of parallelism, large computational requirements and be related with data throughput rather than with the latency of individual operations. Since most ML algorithms and techniques fall under these guidelines, GPUs provide an attractive alternative to the use of dedicated hardware by enabling high performance implementations of ML algorithms. Furthermore, the GPU peak performance is growing at a much faster pace than the CPU performance and since GPUs are used in the large gaming industry, they are mass produced and regularly replaced by new generation with increasing computational power and additional levels of programmability. Consequently, unlike many earlier throughput oriented architectures, they are widely available and relatively inexpensive.

Over the past few years, the GPU has evolved from a special purpose processor for rendering graphics into a highly parallel programmable device that plays an increasing role in scientific computing applications. The benefits of using GPUs for general purpose programming have been recognized for quite some time. Using GPUs for scientific computing allowed a wide range of challenging problems to be solved, providing the mechanisms for researchers to study larger datasets. However, only recently, General Purpose computing on GPU (GPGPU) has become the scientific computing platform of choice, mainly due to the introduction of NVIDIA Compute Unified Device Architecture (CUDA) platform, which allows programmers to use industry standard C language together with extensions to target a general purpose, massively parallel processor (GPU).

The CUDA architecture exposes the GPU as a massive parallel device that operates as a co-processor to the host (CPU). CUDA gives developers access to the virtual instruction set and memory of the parallel computational elements in CUDA GPUs. Using CUDA, the latest Nvidia GPUs become accessible for computation like CPUs. Unlike CPUs, however, GPUs have a parallel throughput architecture that emphasizes executing many concurrent threads slowly, rather than executing a single thread very quickly. This approach of solving general-purpose (i.e., not exclusively graphics) problems on GPUs is known as GPGPU [11].

GPUs are being applied to Intrusion Detection systems deep packet inspection problem for finding several patterns among several independent streams of characters. The highly parallelism of the GPU computation power is used to inspect the packet contents in parallel. Packets of each connection which are in right order compose a stream of characters. Therefore, both levels of parallelism, fine grain and coarse grain, are apparent [10].

The fine grain parallelism, as a fundamental block, is achieved by parallel matching of several patterns against packets of a connection. The coarse grain is achieved by parallel handling of several connections between separate blocks. This approach became possible using CUDA enabled GPUs. A similar approach will be possible using the memristor computer using less energy and increased processing speed.

7 The Field Programmable Gate Array (FPGA) Approach

The third approach to emulating a memristor computer uses a hardware architecture deployed onto a field programmable gate array (FPGA). FPGAs are semiconductor devices that are based around a matrix of configurable logic blocks (CLBs) connected via programmable interconnects. FPGAs can be reprogrammed to desired application or functionality requirements after manufacturing. This feature distinguishes FPGAs from Application Specific Integrated Circuits (ASICs), which are custom manufactured for specific design tasks. Although one-time programmable (OTP) FPGAs are available, the dominant types are SRAM based which can be reprogrammed as the design evolves [12].

Parallelism, modularity and dynamic adaptation are three computational characteristics typically associated with neural networks. FPGA based reconfigurable computing architectures are well suited to implement neural networks as one can exploit concurrency and rapidly reconfigure to adapt the weights and topologies of a neural network. FPGA realization of neural networks with a large number of neurons is still a challenging task because neural network algorithms are "multiplication-rich" and it is relatively expensive to implement.

Usually neural network chips are implemented with neural network trained using software tools in computer system. This makes the neural network chip fixed, with no further training during the use or after fabrication. To overcome this constraint, training algorithm can be implemented in hardware along with the neural network. By doing so, neural chip which is trainable can be implemented.

The limitation in the implementation of neural network on FPGA is the number of multipliers. Even though there is improvement in the FPGA densities, the number of multipliers that needs to be implemented on the FPGA is more for lager and complex neural networks.

The training algorithm is selected mainly considering the hardware perspective. The algorithm should be hardware friendly and should be efficient enough to be implemented along with neural network. This criterion is important because the multipliers present in neural network use most of the FPGA area.

One hardware technique for training is using the back propagation algorithm. The back propagation training algorithm is a supervised learning algorithm for multilayer feed forward neural network. Since it is a supervised learning algorithm, both input and target output vectors are provided for training the network. The error data at the output layer is calculated using network output and target output. Then the error is back propagated to intermediate layers, allowing incoming weights to these layers to be updated [6].

Basically, the error back-propagation process consists of two passes through the different layers of the network: a forward pass and a backward pass. In the forward pass, input vector is applied to the network, and its effect propagates through the network, layer by layer. Finally, a set of outputs is produced as the actual response of the network. The training of the neural network algorithm on the FPGA is implemented using basic digital gates. Basic logic gates form the core of all VLSI design. So, neural network architecture is trained on chip using back propagation algorithm to implement basic logic gates. The architecture of the neural network is implemented using basic digital gates i.e., AND, OR, NAND, NOR, XOR, XNOR function [13].

With the ever-increasing deployment and usage of gigabit networks, traditional network intrusion detection systems (IDSs) have not scaled accordingly. More recently, researchers have been looking at hardware-based solutions that use FPGAs to assist network IDSs, and some proposed systems have been developed that can be scaled to achieve a high speed over 10 Gbps.

FPGA-based implementations of neural networks can be used to detect the system attacks at a high speed and with an acceptable accuracy. Hardware based solution using an FPGA are necessary to fit the high speed performance requirements of modern IDS systems [14, 15, 16]. However, the appropriate choice of the hardware platform is subject to at least two requirements, usually considered independent each other: (1) it needs to be reprogrammable, in order to update the intrusion detection rules each time a new threat arises, and (2) it must be capable of containing the typically very large set of rules of existing NIDSs [12].

Ever increasing deployment of higher link rates and the ever growing Internet traffic volume appears to challenge NIDS solutions purely based on software. Especially, payload inspection (also known as deep packet inspection) appears to be very demanding in terms of processing power, and calls for dedicated hardware systems such as FPGA based systems. FPGA based systems using neural network algorithms takes as input training data to build normal network behavior models. Alarms are raised when any activity deviates from the normal model.

8 Conclusions

In this chapter we covered three approaches to emulate a memristor computer using neural networks, and to demonstrate how a memristor computer could be used to solve Cyber security problems. The memristor emulation neural network approach was divided into three basic deployment methods. The first basic deployments of neural networks are software based algorithms deployed on the traditional Von Neumann CPU architecture (VNCA) using x86 CPUs second, software based algorithms deployed on the VNCA utilizing a Graphics Processing Units (GPUs) and third, a hardware architecture deployed onto a field-programmable gate array (FPGA).

References

1. A. Mitra, W. Najjar, L. Bhuyan, Compiling PCRE to FPGA for accelerating SNORT IDS, in *ACM/IEEE Symposium on Architectures for Networking and Communications Systems*, Orlando, FL, Dec 2007
2. Lower Columbia College, Synapse of interconnecting neurons (2013), http://lowercolumbia. edu/students/academics/facultyPages/rhode-cary/intro-neural-net.htm (Fig. 1 image in chapter) pp. 1223–1230. Accessed 21 Mar 2013
3. Wikipedia, Neuron (2013), http://en.wikipedia.org/wiki/Neuron. Accessed 10 Mar 2013
4. Wikipedia, Neural Network (2013), http://en.wikipedia.org/wiki/Neural_network. Accessed 10 Mar 2013
5. Wikipedia, Feed forward neural network (2013), http://en.wikipedia.org/wiki/ Feedforward_neural_network. Accessed 10 Mar 2013
6. S.L. Pinjare, Implementation of neural network back propagation training algorithm on FPGA, Int. J. Comput. Appl. **52**(6), 0975–8887 (2012)
7. H. Jiang, The application of genetic neural network in network intrusion detection. J. Comput. **4**(12), 1223–1230 (2009)
8. B. Conan, K. Guy, A neural network on GPU (2008), http://www.codeproject.com/Articles/ 24361/A-Neural-Network-on-GPU. Accessed 13 Mar 2008
9. Wikipedia, Neural network software (2013), http://en.wikipedia.org/wiki/Neural_network_ software. Accessed 10 Mar 2013

10. J. Hofmann, *Evolving neural networks on GPUs*, GECCO (2011), http://www.gpgpgpu.com/gecco2011/entries/03/gecco.pdf. Accessed 21 May 2013
11. Wikipedia, Graphics processing unit (2013), http://en.wikipedia.org/wiki/Graphics_processing_unit. Accessed 13 Mar 2013
12. Wikipedia, FPGA (2013), http://en.wikipedia.org/wiki/Field-programmable_gate_array. Accessed 10 Mar 2013
13. P. Lysaght, J. Stockwood, J. Law, D. Girma, Artificial neural network implementation on a fine-grained FPGA, in *Field Programmable Logic and Applications* ed. by Hartenstein, Servít (Springer-Verlag, New York, 1994) pp. 421–431
14. A. Muthuramalingam, S. Himavathi, E. Srinivasan, Neural network implementation using FPGA: issues and application. Int. J. Inf. Commun. Eng. **4**, 6 (2008)
15. Q.A. Tran, Evolving block-based neural network and field programmable gate arrays for host-based intrusion detection system, in 2012 *Fourth International Conference on Knowledge and Systems Engineering* (2012)
16. A.A. Hassan, A. Elnakib, M. Abo-Elsoud, FPGA-based neuro-architecture intrusion detection system, in *International Conference on Computer Engineering & Systems*, Cairo, Egypt, 25–27 Nov 2008, pp. 268–273

Printed in the United States
By Bookmasters